Mushrooms

Mushrooms

A Manual for Cultivation

SUBRATA BISWAS
Senior Scientist (Plant Pathology)
ICAR Research Complex, NEH Region
Tripura Centre, Lembucherra
Tripura

M. DATTA
Joint Director
ICAR Research Complex, NEH Region
Tripura Centre, Lembucherra
Tripura

S.V. NGACHAN
Director
ICAR Research Complex, NEH Region
Umroi Road, Umiam
Meghalaya

PHI Learning Private Limited
Delhi-110092
2025

In fond memory of ***Shri Asoke K. Ghosh*** *(October 1942 – February 2024), Founder Chairman and Managing Director of PHI Learning, whose vision endlessly inspires.*

The Legacy Continues....

Published by Pushpita Ghosh, PHI Learning Private Limited, Rimjhim House, 111, Patparganj Industrial Estate, Delhi-110092 and Printed by Syndicate Binders, A-20, Hosiery Complex, Noida, Phase-II Extension, Noida-201305 (N.C.R. Delhi).

₹595.00

MUSHROOMS—A Manual for Cultivation
Subrata Biswas, M. Datta and S.V. Ngachan

ISBN-978-81-203-4494-5 (Print Book)
ISBN-978-93-5443-247-7 (e-Book)

The export rights of the book are vested solely with the publisher.

Contents

Foreword

Mushroom is an important crop of fungal origin that can be cultivated on several agricultural residues. There are about twenty different mushrooms grown commercially all over the world. These mushrooms are specifically known for their attractive flavours and textures that make a food delicious. They not only contain protein, vitamins and minerals, but also are characterized by low calorie with little fat and very less sugar. They provide a high amount of qualitative nutrition required for our growth and enhance our immune system. The cultivation of mushroom was first started in the middle of seventeenth century in France. Thereafter, it was flourished in different countries. At present, more than hundred countries in the world produce mushrooms commercially. The total global production exceeds 10 million tons and is valued at US$ 25 billion per annum. On the contrary, in India, mushroom cultivation was introduced very late, during the middle of twentieth century, but it flourished very rapidly and extended to almost all parts of the country. Now, our country produces about 1 lakh metric ton of fresh mushroom every year. Four different mushrooms viz. button mushroom (*Agaricus bisporus*), oyster mushroom (*Pleurotus sajor-caju*), paddy straw mushroom (*Volvariella volvacea*) and milky mushroom (*Calocybe indica*), have been brought under commercial cultivation. In addition, several other mushrooms have been identified as promising material for commercial cultivation.

In this book, '*Mushrooms—A Manual for Cultivation*', the authors have given emphasis on describing the different kinds of commercially important mushrooms. They have explained their structure, natural diversity, food and medicinal values, impact of climatic factors on their cultivation and cultivation methodologies starting from the laboratory set up to post harvest technologies for developing a complete package and practices of different mushroom cultivations. The economics of mushroom cultivation and ancillary information about mushroom centres, sources of spawn and machineries as well as addresses of leading mushroom farms and exporters have been given in this book. All information provided here will be very useful to start this venture. In this book the authors have synchronized both theoretical and practical knowledge and generated new and useful information required for students, researchers, teachers, mushroom growers and exporters. I am confident that this book will serve as a handy compendium for all those who are involved in mushroom development programmes in our country.

Dr. Anil Kumar Singh
DEPUTY DIRECTOR GENERAL (NRM)
INDIAN COUNCIL OF AGRICULTURAL RESEARCH
NEW DELHI

Preface

Mushroom is considered as an alternate food for the increasing population of the present era in the world. The term *mushroom* refers to a group of fungi having large sporophores. It may be any member of higher fungi comprising of both edible and poisonous ones. In this book only the edible species of mushroom have been discussed. There are about 2000 fungal species all over the world, which are edible. These edible species are recognized as delicious foods of good quality protein. These are also rich in vitamins and minerals. Consumption of these mushrooms as food provides us high amount of qualitative nutrition and different medicines to combat diseases and to increase immune systems in our body. In spite of that, mushroom is not generally consumed for its food and medicinal values. It is mostly consumed for its unique and tempting flavour and texture, that make a food delicious. Further, the presence of mushroom dish on any occasion or in a hotel menu represents the status symbol.

The cultivation of mushroom was first started in France sometimes during the seventeenth century. Thereafter, its cultivation was flourished to different European, North American and Southeast Asian countries. In India, the mushroom cultivation, in true sense, was started during the year 1961 with the launch of a scheme of ICAR at Solan in Himachal Pradesh. At present, the global production of mushroom, which mostly represents the production of only 7–8 mushroom species, exceeds 10 million tons that valued above 25 billion dollars. In India, on the other hand, the total annual production of all kinds of mushroom is at best 1 lakh metric ton. Mushroom is a useful recycler of the waste materials, particularly of the agricultural residues. A large variety of agricultural residues which are not suitable as cattle fodder may also be used for the mushroom cultivation. Growing mushrooms on agricultural residues is a traditional practice, which is followed even today with certain modifications and improvement.

In the present book, the edible species which are commercially grown or having potentials to grow on commercial scale in India are considered for discussion. The book comprises fourteen chapters, of which the first four chapters are the introductory ones and deal with the general features of mushrooms, their diversity and production status in the world, influence of climatic factors in production and their food and medicinal values. The chapters five to eight, which are the main focus of our book deals with the laboratory and farm set up procedures, different technologies of spawn preparation and detailed estimates of various mushrooms giving their morphology, cultivation technologies, and management of diseases and pests including weeds and fungi. The ninth chapter gives the methodologies of post harvest technologies for storage, food preparations and preservations of mushrooms, while, the tenth chapter deals with the economics of mushroom production and the eleventh chapter deals with ancillary information on the sources of different equipments, machineries and mushroom cultures and the addresses of mushroom farms and exporters for helping people establish mushroom farms. The last three chapters are the concluding ones that present references and glossary terms.

In writing this book, we have synchronized both practical and theoretical knowledge into a common platform to generate new and important information required for the students, researchers, teachers, mushroom growers and exporters. We have consulted several leading books, scientific journals, bulletins, training manuals, reports and Internet documents before writing this book, apart from the *in situ* practical experiences and finally formulated the information into our own way of presentation. We gratefully acknowledge the authors whose works are referenced in this book. We are also thankful to our earlier Director, Dr. K.M. Bujarbaruah, of ICAR Research Complex for NEH Region, Umiam and earlier Joint Director, Dr. N.P. Singh, of this centre for their continuous help and encouragement in establishing a mushroom laboratory and conducting various researches in this line. Since, without that, the information which included in this book would not be possible to generate. The authors are also thankful to all staff members of Plant Pathology Division, ICAR Research Complex for NEH Region, Tripura Centre, Lembucherra, Tripura for their assistance in mushroom cultivation and research which eventually helps in writing this manuscript.

Subrata Biswas
M. Datta
S.V. Ngachan

CHAPTER

1 Introduction

Mushrooms are a group of fungi having large sporophores. They occur seasonally all over the world in various habitats varying from sandy plains to thick forests or green meadows to roadside pathways. The mushrooms are of various shapes, sizes, colours and tastes. Several of them are edible while others (nearly 100 species) are non-edible (non-palatable) or poisonous, even deadly poisonous (*Amanita phaloides, A. virosa, Cortinarius rubellus, C. orellanus*, etc.). The consumption of mushroom from its wild habitat needs adequate precautions to avoid accidental death due to mushroom poisoning. It is estimated that about 2000 fungal species are edible all over the world of which 300 species belonging to 70 genera are reported from India (Chadha and Sharma 1995). So far, only a few (20) have been brought under commercial cultivation and 4 to 5 are under industrial scale of production in about 100 countries (Chang and Miles 1991). Total mushroom production in the world has been increased more than 18-folds from about 350,000 metric tons in 1965 to about 6,160,800 metric tons in 1997. The bulk of this increase has occurred during the last 15 years. A considerable shift has been occurred in the composite of genera that constitute the mushroom supply. During the 1979 production year, the button mushroom, *Agaricus bisporus*, accounted for over 70 per cent of the world's supply. By 1997, only 32 per cent of world production was of *A. bisporus*. The People's Republic of China is the major producer of edible mushrooms, producing about 3,918,300 tons each year or about 64 per cent of the world's total (Royse 2003). The major mushroom growing countries in the world are China, Germany, the Netherlands, Canada, France, UK, Sweden, USA, Italy, Taiwan, Korea and Indonesia. Seven genera, viz. *Agaricus*, *Lentinus*, *Volvariella*, *Pleurotus*, *Auricularia*, *Flammulina* and *Tremella* contribute about 89 per cent of the total world production. The mushroom production in Western countries is mostly dominated by the production of button mushroom (*Agaricus bisporus*), while, that in East Asian countries is dominated by specialty mushrooms (Chang 2007). According to the estimate based on the production year 2003–04, USA alone produced 381,479.4 metric tons button mushrooms which share 98.4 per cent of total mushroom production of 387,534.7 metric tons in the world. In Spain, which is the third largest mushroom producing country in Europe, the total mushroom production is 110,000 tons (in 2004) out of which 80 per cent is of button mushroom, 15 per cent of oyster and 5 per cent of shiitake mushrooms. In 2003 in East Asian countries, the button mushroom (*Agaricus* sp.) was produced 1,330,000 metric tons in China, 0 metric ton in Japan, 19,790 metric tons in South Korea and 4276 metric tons in Taiwan, sharing 12.8 per cent, 0 per cent, 11.6 per cent and 4 per cent of the total production of their respective countries (Chang 2006, Cui 2004, Ho and Peng 2006, Yamanaka 2006). The production of oyster mushroom is highest 6,046,400 metric tons/year, sharing 58.2 per cent of total mushroom production in China, followed by South Korea with 123,631 metric tons which shared 72.7 per cent of total mushroom production in the country. The shiitake mushroom is mainly produced in China (with 2,228,000 metric tons yearly production), South Korea (41,876 metric tons/year), Taiwan (36,000 metric tons) and Japan (35,294 metric tons).

In India the possibilities for establishment of mushroom industry was first visualized by Professor S.R. Bose in 1921, with the knowledge of cultivation of two agarics on sterilized dung medium. The paddy straw mushroom was first cultivated by Thomas et al. (1943) in Coimbatore, while oyster mushroom by Bano et al. (1962). The cultivation of button mushroom (*Agaricus bisporus*) has been started systematically during the year 1961 with the launch of a scheme on the 'development of mushroom cultivation in Himachal Pradesh' by ICAR at Solan, Himachal Pradesh. Till 1980, its cultivation was mainly confined to the northern hill states of India, like Himachal Pradesh and Jammu and Kashmir. Thereafter, remarkable change has been occurred in button mushroom scenario with the spread of its cultivation to the East, in West Bengal and North East Hill (NEH) region (Chadha 1992). At present, the total button mushroom production in India is estimated about 50,000 tons (Verma 2002a). This is comprised 85 per cent of the total mushroom production in India. The mushroom production in Himachal Pradesh is more than 8000 tons per year where two export oriented farms at Paonta Sahib and Nalagarh, each producing 3500–4000 tons per year (Suman and Sharma 2005). In Punjab, several cold storages are modified to mushroom production unit and one such unit at Lalru produces about 3500–4000 tons mushroom per year (Dhar 1997). In Haryana total mushroom production is about 8000 tons of which 75 per cent come from seasonal growing. The seasonal growers with the use of unpasteurized compost and mushroom production capacity 80–100 kg per ton compost are the major contributors of total mushroom production in India. The button mushroom production has increased much during the last two decades and that is evident from the export scenario of India. In 1991, the country had exported only 790 kg processed mushroom which by 2001–02 increased to 11.8 million kg. In 2008–09, the country had exported 15.1 million kg processed mushroom and 0.06 million kg fresh mushroom. However, due to some other problems, like global recession and increase in local consumption, in the recent past (2009–10) the export of processed mushroom has declined to 9.3 million kg, although, that of fresh mushroom is increased to 0.39 million kg (Singh et at. 2011).

Apart from button mushroom, three other mushrooms, like oyster mushroom (*Pleurotus sajor-caju*), paddy straw mushroom (*Volvariella volvacea*) and milky mushroom (*Calocybe indica*) are commercially cultivated in India. These mushrooms are grown in small farms or in farmer's houses as cottage industry. Amongst the three, oyster mushroom is the most important and gaining much popularity during the days, due to its easy method of cultivation and ability to grow on various waste materials with high production efficiency. This mushroom is suitable to grow in tropical and subtropical areas. Paddy straw mushroom due to its low and inconsistent yield has not been popularized much for commercial growing, although, it has the potential for industrialization under tropical climatic condition with the use of partially decomposed compost and canning the harvested products. At present, its cultivation is mainly confined to Orissa and West Bengal. The cultivation of milky mushroom is of recent origin (Purkayastha and Nayak 1981a), but it gained much popularity under tropical climatic conditions. It is now cultivated in different states of South India and Rajasthan .

The cultivation of mushroom was first started with the white button mushroom in France sometimes during 1630 (Atkins 1983). Bonnefons (1650) gave the first account on the cultivation of mushrooms by the melon growers of Paris (France) who started growing the cultivated mushroom in open. Tourneforte (1707) described the first method of mushroom cultivation and the use of casing soil. Lundenberg (1754) described the conditions of open air growing. Chambry (1810), a French gardener, showed that underground quarries without light favoured the growth of mushrooms. Around 1825, in the Netherlands, mushroom cultivation was started in caves and later that was followed in France during 1848 and successively in other Europian countries. In cave, the mushroom cultivation was done initially with the use of composted horse manure and pregerminated mushroom spores or spawn. A casing material made up of rotten leaf litter was later added to stimulate mushroom formation. Callow (1831) showed that mushroom production was possible all round the year in rooms in England. However, the commercial mushroom cultivation was started following the preparation of 'standard mushroom house' in the United States, and large scale culture in caves around Vicenza in Italy during the years, 1910 and 1913, respectively. This was flourished in different countries after the Second World War (Delmas 1978, Sohi 1988).

Mushrooms, being fungi, are heterotrophic in nature, they cannot prepare their own food. They have no chlorophyll to transform sunlight energy to chemical energy. However, they have the ability to convert the complex lignocellulosic waste materials into soluble absorbable substances for their nourishment. The mycelia (vegetative structures) of mushrooms perform this job by producing extra cellular enzymes which digest the complex carbohydrate, lignin, lipids and proteins into a form of simple molecules that are easily absorbed by them. During the course of development, the mycelia enter into reproductive phase by producing sporophores or sporocarps, which we know as mushrooms.

As regards its morphology, mushroom has three distinct parts, base, stipe and cap. However, sometimes certain parts are not formed, depending upon the types of sporophores which either may be ascocarp or basidiocarp. The mushrooms belonging to Ascomycotina form ascocarps, which may be of different sizes and shapes and contain a hymenial layer for bearing ascospores. The ascocarp may be cup shaped, bearing special sponge like structure as in *Morchella* sp., bell shaped, saddle shaped or closed fruit bodies of truffles (*Tuber* sp.). In contrast, the mushrooms belonging to Basidiomycotina form basidiocarps. The basidiocarps may be stipitate, i.e. with stipe, substipitate or sessile. Pileus or cap may be infundibuliform, convex, campanulate, flabelliform, umbonate, umblicate, conical, applanate, petaloid, gibbous, involute, reflexed, etc. The margins of pileus may be regular or irregular it may contain veil or not. The lower side of the cap is composed of gills (in case agarics) or pores (bolete) containing hymenial layer with basidia and basidiospores. The gills may be free or attached with stipe. The stipe may be hollow or solid. It may be of different sizes with or without a ring (annulus). The base may be swollen bulbous or not. It may be attached with universal volva or not (Purkayastha and Chandra 1985).

Unlike green plants, almost all the mushrooms are haploid. The diploid phase is normally transient and restricted to hymenium, more precisely in ascogonium and basidium, where meiosis occurs to form spores. Nuclear behaviour in mushrooms is not always clear-cut, since, the hyphal cell may or may not contain same number of nuclei thus by nature they are heterokaryotic. In the hetrokaryotic mushrooms, each nucleus is independent, but the structure and behaviour of the mushrooms are controlled by the kinds of genes and the proportion of each kind of genes, regardless of whether these are separated into different nuclei or not (Alexopoulos et al., 2002). In mushroom, like *Volvariella volvacea*, the number of nuclei vary from 1 to 15, while, in *Lentinula edodes* and *Pleurotus ostreatus,* it is 1 or 2. The structure of nucleus is either diffused or constricted. The nuclear migration generally occurs during hyphal anastomosis (Chang 1978).

Mushrooms are recognized as the alternate source of good quality protein and these are capable to produce highest quantity of protein per unit area and time from the worthless agro wastes. These are also the good source of vitamins and minerals with certain medicinal properties. Thus, mushroom production is nowadays very much important among biotechnical practices all over the world with the use of different methodologies as suited with the place and mushroom species. However, before adopting any given technology, adequate knowledge about the geographical position, climatic conditions, types of inhabitants, their population and choice of food habits, major crops and plantations of the site are essential in order to understand available agro-wastes, mushroom species suitable for growing, target group of mushroom consumption and the marketing facilities (Biswas and Singh 2008).

CHAPTER 2

Biodiversity of Mushrooms

Mushrooms are a group of fungi having large sporophores. They belong to two different subdivisions, Ascomycotina and Basidiomycotina of fungi. The mushrooms belonging to Ascomycotina form ascocarps, which may be of different sizes and shapes. The ascocarp consists hymenial layer for bearing ascospores in ascus. In Ascomycotina, the most important mushroom is *Morchella* (morel) that belongs to the order Pezizales under the group of Discomycetes (cup fungi) in class Ascomycetes. The ascocarp of morel is club shaped and the upper part of the club is pitted or ridged. The club consists of thin shell of tissue around a hollow insertion. Thus, there is relatively little substance to it. All true morels, like conic morel (*Morchella conica*), the hybrid morel (*Morchella semilibera = M. hybrida*), delicious morel (*Morchella deliciosa*), the common morel (*Morchella esculenta*) and the thick stemmed morel (*Morchella crassipes*) are edible and delicious (Alexopoulos and Mims 1979). The common morel (*M. esculenta*) and the delicious morel (*M. deliciosa*) are very frequent in India particularly the former one in upper part of Uttar Pradesh (UP), Himachal Pradesh (HP), Punjab and Kashmir (Sharma 2004). This mushroom is easy to culture in laboratory, but very difficult for fruit body formation. Another important member in this group is *Tube*r (also known as truffle), belonging to the order Tuberales. The ascocarp of *Tuber* is a closed cup. Interior of ascocarp is filled up with folds and veins growing inward to the shell (periderm). The fruit body is irregularly globose, fleshy or woody and has resemblance to a small potato. The truffles (*Tuber*) are mycorrhizal fungi; always grow in association with trees in the forest. Thus, these are very difficult to grow artificially in houses. In most cases, these are used as food collected only from the wild habitat. The place where these are grown in nature is very difficult to detect, since, the ascocarps developed in soil rarely become exposed over the upper surface. In many cases, these mushrooms are commercially collected from the soil with the help of trained dogs or pigs, who can identify the mature truffle mushrooms from their odour. The species *Tuber aestivum*, *T. uncinatum*, *T. melanosporum*, *T. mesentericum, T. monotanum and T. brummale*, etc. are important truffle mushrooms.

The mushrooms belonging to Basidiomycotina form basidiocarps. The basidiocarp has three distinct parts, base, stipe and cap. However, sometimes certain parts are not formed or formed as rudimentary. The basidiocarps may be stipitate, i.e. with stipe, substipitate or sessile. Pileus or cap may be infundibuliform, convex, campanulate, flabelliform, umbonate, umblicate, conical, applanate, petaloid, gibbous, involute, reflexed, etc. The margins of pileus may be regular or irregular it may contain veil or not. The lower side of the cap is composed of gills (in case agarics) or of pores (bolete) containing hymenial layer with basidia and basidiospores. The gills may be free or attached with stipe. The stipe may be hollow or solid and of different sizes with or without a ring (annulus). The base may be swollen bulbous or not. It may be attached with universal volva or not (Purkayastha and Chandra 1985).

The edible mushrooms in this group mostly fall under the order Agaricales. The two important members of basidiomycotina, *Auricularia* and *Tremella* are considered first, ahead of the main group Agaricales, due to their saucer-like to irregular fruit body with cartilaginous to jelly like consistency (Smith 1978). These two genera are included in the subclass, Phragmobasidiomycetidae (Ainsworth et al. 1973). Several species of *Auricularia* under the order Auriculariales are edible. The most important ones are *Auricularia auricula* (Mou-Erh in Chinese) and *A. polytricha* (Mouh Leh in the Orient). The former one is generally collected from the local habitats and the latter one is cultivated. In nature, they grow on wood logs. The members of this genus are characterized by gelatinous, irregular and foliose fruit bodies. The smooth surface of the basidiocarp is covered by the hymenium. The trama is mostly made up of gelatinous hyphae which cause the fruit body to have a tough cartilaginous consistency. The genus *Tremella* is under the order Tremellales. The members of this order are commonly known as jelly fungi because of their gelatinous, jelly-like nature of basidiocarps. The fruiting body of *Tremella fuciformis* is pure white. It consists of a number of large leaf-like folds on which the hymenium is formed. This mushroom is used as food, also considered as medicinal mushroom and cultivated on commercial basis in many countries, especially in East Asian countries. As the mushroom has jelly-like consistency, it dries down and reduces to almost nothing, but can readily revived by in soaking water. It is marketed fresh near the point of production or preserved in dry form for selling afterwards to places far from the point of production.

The other mushrooms which are commonly cultivated or used as food are under the main group Agaricales. The term 'mushroom' is mainly referred to the fungi of this group due to appearance of umbrella-like sporophore with cap, stipe and base, and bearing basidia and basidiophores on the surface of gills, except boletes which are a group of fungi also included in this order. In Agaricales, most of the edible fungi belong to the families Agaricaceae, Lepiotaceae, Russulaceae, Boletaceae, Strophariaceae, Pluteaceae and Tricholomataceae. The sporophores of boletes resemble to typical mushroom, but the basidia and spores are formed in deep tubular pores or shallow pits. In boletes the most important edible species is *Boletus edulis* under the family Boletaceae (Garcha 1980, Purkayastha and Chandra 1985). The sporophore of this mushroom is grown as white with slightly viscid glabrous surface. The central stem (stipe) is broad bulbous and unlike other species of this genus, no colour change occurs in flesh on cutting. This mushroom grows in forest or in gardens as mycorrhizal association with trees. Several other mycorrhizal species of genera *Russula* and *Lactarius* under the family Russulaceae in Agaricales are edible. The members of this family are characterized by the presence of sphaerocysts in the context of pileus and often in trama. The species of *Russula* are easily recognized by the presence of soft colourful caps with very brittle gills and short thick stipes. The important edible species in this genus are *Russula cyanoxantha* and *R. delica* (milky white cap), and *R. virescens* (Delmas 1978, Jana and Purkayastha 1982). The species of *Lactarius* exude a watery or milky juice if a cut is made in the fresh. This juice flows in the lactiferous tubes and it may be colourful (creamy, yellow, blue or red or colourless). The species *Lactarius sanguifluus* (sanguin) and *L. deliciosus* (with saffron milk cap) are good edible mushrooms sold in different markets in France (Delmas 1978).

In the family Tricholomataceae a large number of fungi are edible. It is composed of the members with white spores and attached gills. The oyster mushrooms (*Pleurotus* sp.) are classified in this group. These are characterized by their oyster shell-like caps that appear on logs or tree stumps in shelf-like layers. Several species, like *Pleurotus sajor-caju, P. flabellatus,, P. florida, P. ostreatus, P. citrinopileatus, P. cornucopiae, P. sapidus, P. membranaceous, P. eryngii, P. fosssulatus, P. eous, P. djamor and P. platypus* are brought under artificial cultivation. They are very good to eat and possess good flavour. Apart from this, several other species of this genus are consumed by collecting from the local habitats, viz., *Pleurouts squarrosulus, P. dryinus*, etc. (Purkayastha and Chandra 1985). The genus *Tricholoma* of this family also has some good edible species. The species *Tricholoma matsutake* and *T. terreum* (*petit gris*) are very important edible members in this genus which grow as mycorrhizal fungi in coniferous forests (Tominaga 1978, Watling and Gregory 1980). The shiitake mushroom *Lentinula edodes* (=*Lentinus edodes*) under this family is drawing most attention of the growers and consumers in the world for its enormous medicinal and food values. This mushroom usually grows on wood of dead deciduous trees of *Fagales* Basidiocarp of this mushroom is usually eccentric or centrally stipitate, arising solitary or in small groups. The cap is pale to dark reddish brown, dry and more or less fibrillose, the cuticle breaking into scales of various sizes and shapes, often rimose with white context showing. Another important member in this family is the winter mushroom *Flammulina velutipes*, which is cultivated commercially all over the world. This mushroom is also known as golden needle mushroom, enoki, enokitake, velvet foot or velvet stem mushroom at different places. This winter mushroom is particularly known for its taste and medicinal properties. In nature, the small very viscid orange coloured basidiocarp of this mushroom grows in cluster clumps on dead and decaying wood during winter. Several species of the genus *Termitomyces* are very good edible mushroom. The mushrooms have association with the termites and these generally grow on sandy places where termite nests are present. The species, *Termitomyces albuminosa, T. cartilaginous, T. clypeatus, T. eurhizus, T. heimii, T. mammiformis* and *T. radicatus*, all are important edible mushrooms, generally sold in market, collected from their wild habitats. All these mushrooms have pseudorrhiza at the base of stipe.

In Lepiotaceae, several species of *Lepiota* are good edible fungi, which are superficially resembled to certain species of *Amanita* with the presence of white spores, free gills and an annulus. However, unlike *Amanitas,* the members of this family have no volva. The most important member of edible fungi in this family is *Macrolepiota procera* (=*Lepiota procera*), popularly known as parasol mushroom. In nature this mushroom grows solitary or in small clusters on soil in pastures, lawns, woods, gardens or road side. This mushroom is of pale coloured cap. The cap is ovate at young, later campanulate, and finally expanded with a brown umbo. Surface of pileus is cracking into large reddish brown scales arranged concentrically with white small scales scattered in between the large ones. The stipe is long cylindrical with bulbous base with thick tough annulus, but devoid of volva. Basidiospores are white (Delmas 1978, Purkayastha and Chandra 1985, Biswas 1988). This edible species also has some resemblance with the green spored poisonous species *Chlorophyllum molydites* (Alexopoulos and Mims 1979).

The members of Pluteaceae produce pink spores. The important edible mushrooms in this family are *Pluteus cervinus* that grows on sawdust and *Volvariella volvacea* that grows on paddy straw and commercially cultivated as paddy straw mushroom. The former one is characterized with its white, yellowish or brown sporocarp that devoid volva and of poor quality in edibility, while, the latter one with its grayish sporocarp that has universal volva and of excellent quality in edibility. Several species of *Volvariella* are edible. In India, although reported that three species, like *V. volvacea V. diplasia* and *V. esculenta* are cultivated (Roy et al. 1978). However, the separate entity of the species is doubtful due to absence of noticeable antigenic differences in the precipitin tests and presence of similarity in morphological characteristics among the cultivated strains. Thus, it is concluded that all cultivated forms of paddy straw mushroom in the world are the members of *V. volvacea* (Singer 1975, Chang 1978a).

The most important mushroom in the world is under the family Agaricaceae. The common mushroom or white button mushroom *Agaricus bisporus* and the temperature tolerant white button mushroom *Agaricus bitorquis*, both are the members in this family. The mushroom appears as white to brown button shaped cap with free gills, an annulus, no volva, and a central stalk that readily separates from the cap (Alexopoulos et al. 2002).

In family Strophariaceae, the most important mushroom is the giant mushroom *Stropharia rugoso-annulata*, is cultivated on cereal straw without any admixtures of an organic origin or of an inorganic nature (Szudyga 1978). The pileus of this mushroom is of 5–40 cm broad at young stage that is white, but later it turns to yellow or brown with reddish tinge. The gills of this mushroom are attached and spores are dark purple brown. This mushroom is very good in edibility and easy to cultivate, but it has some drawbacks to grow commercially. These drawbacks are the unstable yield, not suited for intensive raising, very brittle fruit bodies, hollow stem and limited keeping quality due to regrowth of mycelia from its fruit body (Szudyga 1978).

In Agaricales, apart from the above mentioned families, other families also contribute a few good edible mushrooms. For example, in Amanitaceae, the well known white spored genus *Amanita* has a few edible species like *A. vaginata*, *A. rubescens* and *A. caesarea* which is generally collected from forest from their mycorrhizal association and marketed in different markets in many countries including India (Delmas 1978, Purkayastha and Chandra 1985, Biswas 1988). The members of this genus are characterized by the presence of free gills, annulus and volva (Alexopoulos and Mims 1979). However, the use of these edible mushrooms as food needs special cautions, since, several species of this genus, like *Amanita phalloides*, *A. virosa*, *A. verna*, *A. pantherina* and *A. muscaria* are deadly poisonous and the mushrooms possessing both volva and annulas are difficult to identify the correct edible species in nature. In family Cortinariaceae, the genus *Pholiota* also contributes a number of edible species. Amongst them, *Pholiota nameko* which is known as nameko (=viscid mushroom) in Japan, is very important wood inhabiting cultivated mushroom. In nature it grows on dead trunks or stumps of deciduous trees, especially of Fagaceae. The pale brown to bay brown coloured cap of sporophore is hemispherical to convex with smooth glabrous surface covered with glutinous mucilage. This mushroom is commercially cultivated on sawdust medium supplemented with rice bran in Japan and other East Asian countries (Arita 1978). In family Coprinaceae, the edible mushroom *Coprinus fimetarius* which has excellent flavour and good shelf life under refrigerated condition, is one of the important cultivated mushroom species (Kurtzman 1978). This mushroom is generally cultivated on cereal straw supplemented with calcium nitrate.

Amongst the other mushrooms, the mycorrhizal species *Cantharellus cibarius* of family Cantharellaceae under the order Aphyllophorales is of great importance for its good edibility (Delmas 1978). The mushroom has the cap that taperes towards the stalk forming a deeply depressed centre at pileus that appears as funnel. The mushrooms are generally sold in different markets after collecting them from their mycorrhizal wild habitats.

In other way, the edible mushrooms are also classified on the basis of substrate requirements into five groups as follows (Bels 1976):

Group 1: Those which grow on fresh or almost fresh plant residues,

e.g. *Lentinus, Flammulina, Auricularia, Pholiota, Tremella, Pleurotus.*

Group 2: Those which grow on slightly composted material,

e.g. *Volvariella, Stropharia, Coprinus.*

Group 3: Those which grow on very well composted material,

e.g. *Agaricus.*

Group 4: Those which grow on soil and humus,

e.g. *Lepiota, Morchella.*

Group 5: Mycorrhizal fungi,

e.g. *Tuber, Morchella, Lactarius, Amanita, Boletus, Cantharellus.*

Unlike green plants, almost all the mushrooms are haploid. The diploid phase is normally transient and restricted to hymenium, more precisely in ascogonium and basidium, where meiosis occurs to form spores. Nuclear behaviour in mushrooms is not always clear-cut, since, the hyphal cell may or may not contain same number of nuclei, thus, by nature they are heterokaryotic. In the hetrokaryotic mushrooms, each nucleus is independent, but the structure and behaviour of the mushrooms are controlled by the kinds of genes and the proportion of each kind of genes that contains, regardless of whether these are separated into different nuclei or not (Alexopoulos et al. 2002). In mushrooms, like *Volvariella volvacea*, the number of nuclei vary from 1 to 15, while, in *Lentinula edodes* and *Pleurotus ostreatus* it is 1 or 2. The structure of nucleus is either diffused or constricted. The nuclear migration generally occurs during hyphal anastomosis (Chang 1978).

CHAPTER

3

Influence of Climatic Factors

Edible mushrooms can survive in a wide range of climatic conditions, but for satisfactory commercial production they must be given conditions fairly near to the ideal at each successive stage of growth. Failure to do this will result in weaker or slower growth. A large number of climatic factors are involved in the growth and development of mushrooms. In the current treatise, a few of the major ones such as temperature, humidity, rainfall, light and CO_2 contents in air have been discussed.

TEMPERATURE

Mushrooms grow in a wide range of temperature. Some show preference for low temperature (*Agaricus bisporus, Lentinula edodes, Pleurotus* sp.) while the others require higher temperatures (*Calocybe indica, Volvariella volvacea*) for their growth and development. Therefore, the optimum temperature requirement of any given mushroom should be ascertained before it is selected for commercial cultivation in a particular place. Further, it is necessary to keep in mind that different strains of same species occasionally differ markedly in their response to temperature. Some of the important cultivated mushrooms and their optimal temperature requirements are presented in Table 3.1. Kim (1978) has reported that spring and autumn seasons of Korea having a temperature range of 10–20ºC are the best natural conditions for growing button mushroom (*A. bisporus*). However, in Indian conditions, 16–18ºC is considered as the most suitable temperature for fructification of the mushroom, although its mycelial growth and spawn run require to some extent higher temperature 22–25ºC (Sohi 1988, Bahukhandi and Sharma 2002). Apart from this, a high temperature-tolerant button mushroom (*A. bitorquis*) has also been introduced for cultivation under the tropical and sub-tropical climatic conditions in India due to infeasibility of the common button mushroom (*A. bisporus*) for cultivation. As regards the milky/*dudh* mushroom, it grows well during summer and rainy season in Tripura (Biswas and Singh 2009a). This mushroom requires higher temperature (30–32ºC) for fructification than that required for mycelial growth which is in contrast to other edible mushrooms (Sharma et al. 2006). Fluctuation of temperature is very much essential to induce

Table 3.1 Temperature Requirement for Optimal Growth of Different Cultivated Mushrooms

Mushroom	*Growth Stage*	*Temperature Requirement* (ºC)	*Source*
Agaricus bisporus (button mushroom)	Mycelial growth	26.7	Dugger (1901)
		25	Lambert (1932)
		24	Treschow (1944)
		24–25	Hayes (1978)
		22–25	Bahukhandi and Sharma (2002)
	Spawn growth	20–25	Edwards (1978)
	Fructification	20–22	Edwards (1978)
		19.44 (67ºF) –21.66 (71ºF)	Samp and Phelps (1986)
		16–18	Sohi (1988)
A. bitroquis (button mushroom)	Spawn growth	29–31	Chadha and Sharma (1995)
	Fructification	25–26	Chadha and Sharma (1995)
Calocybe indica (*dudh/milky* mushroom)	Mycelial growth	25	Sharma et al. (2006)
	Spawn growth	27–28	Sharma et al. (2006)
	Fructification	30–32	Sharma et al. (2006)
Lentinula edodes (shiitake mushroom)	Mycelial growth	25	Tokimoto and Kosmatsu (1978), Sharma et al. (2006)
		24–28	Ito (1978)
	Fructification	12–20	Ito (1978)
		23–25	Sharma et al. (2006)
Pleurotus eous (oyster mushroom)	Fructification	21–35	Purkayastha and Chandra (1985)
P. eryngie (oyster mushroom)	Mycelial growth	25	Zadrazil (1978)
		25	Kong (2004)
	Fructification	13–18	Kong (2004)
P. flabellatus (oyster mushroom)	Fructification	20–30	Suman and Sharma (1999)
P. florida (oyster mushroom)	Mycelial growth	30	Zadrazil (1978)
		25	Kong (2004)
	Fructification	15–25	Kong (2004)
P. ostreatus (oyster mushroom)	Mycelial growth	30	Zadrazil (1978)
		25	Kong (2004)
	Fructification	10–17	Kong (2004)
P. sajor–caju (oyster mushroom)	Mycelial growth	22–28	Bahukhandi and Sharma (2002)
		25	Kong (2004)
	Fructification	22–30	Charkarvarty and Sarkar (1982)
		22–28	Bahukhandi and Sharma (2002)
		23.6–28.5	Biswas and Singh (2007)
		18–25	Kong (2004)
Volvariella volvacea (paddy straw mushroom)	Mycelial growth	35	Chang and Chu (1969), Biswas (1988)
		30	Chandra and Purkayastha (1977)
	Spawn growth	34–35	Chang (1978)
	Fructification	28–32	Chang (1978)
		30–43.2	Purkayastha et al. (1981b)
		30–42	Suman and Sharma (1999)

mycelial differentiation in shiitake mushroom and king oyster mushroom (Komatsu 1961, Kong 2004). Thus, in certain cases, chilling treatment with ice is in practice for fruit body initiation. Ito (1978) has suggested maintaining 12–20ºC for fructification of shiitake mushroom. Oyster mushrooms (*Pleurotus* sp.) are mostly cultivated within the temperature range 20–30ºC. The growing season of oyster mushrooms in north India is October to April, while in north east hills during September to mid November and in Tripura throughout the year (Bahukhandi and Sharma 2002, Chandra et al. 1998, Biswas and Singh 2007). The range of maximum temperature 20–28ºC and the minimum 12.2–20.5ºC has been determined as the most suitable period for *P. sajor-caju* (Chandra et al. 1998). Biswas and Singh (2007) have reported that the maximum temperature of atmosphere has greatly influenced on and that correlates negatively with the yield. The spores of *V. volvacea* germinate well at 40ºC, but the initial hyphal growth requires 35ºC (Chang and Chu 1969). The optimum temperature for the growth of mycelia is 30–35ºC and the growth rate drops markedly at 20 and 40ºC (Chang 1978b, Biswas 1988). Further, it is to note that the production of paddy straw mushroom is quite sensitive to environmental factors. Purkayastha et al. (1980, 1981a, b) have found that June and July in West Bengal with temperature range 22.3–43.2ºC are the most favourable months for paddy straw mushroom production, although, the air temperature 28–32ºC is the most suitable condition for fruit body formation and further increase in temperature though enhances the fruit body development, but reduces the yield and quality of the mushroom (Chang 1978a).

RELATIVE HUMIDITY

High relative humidity is favourable for growing all mushrooms. It also prevents drying up of compost and substrate surfaces. Sohi (1988) has suggested maintaining 80–90 per cent relative humidity with moisture content of 62–63 per cent (in plastic bag method) and 65–67 per cent (in tray method) in compost for button mushroom cultivation. Moisture content of straw as well as relative humidity has been found to play certain important roles in *V. volvacea* cultivation. Under natural conditions bed moisture 65–70 per cent and the relative humidity 80–90 per cent around the beds should be maintained to cultivate that mushroom (Chang 1965, Yau and Chang 1972). High moisture content (90–95 per cent) is also required for *C. indica* cultivation (Purkayastha and Chandra 1985). In case of oyster mushroom, high humidity (80 per cent) is maintained by hanging wet gunny cloths inside the mushroom house for initiation of fruit bodies and their developments (Chandra et al. 1995).

RAINFALL

So far, rainfall has no direct impact in mushroom cultivation under in-house condition, except that it causes rising of relative humidity to a great extent which favours fruit body formation and its growth. On the other hand, in nature its effect is enormous. It moistens substrates and creates the conditions favourable for the mushrooms' growth. Practically, almost all the cultivated mushrooms of India grow on damp soil rich in organic matter or on rotten wood of trees, etc. after rain or in rainy season (Purkayastha and Chandra 1985).

LIGHT

Light is also an important factor for initiation of fruit bodies. In *Pleurotus* sp., exposure to light (artificial or sunlight) is necessary for initiation of fruit body primordia, and thus, light is given for at least 15 minutes per day (Zadrazil 1978). There is, also a positive phototropism in *Pleurotus* species during the development of fruit bodies (Block et al. 1959, Gyurko 1972). Stimulation and retardation of mycelial growth in light (2000 lux) are also evident with *V. volvacea* and *Macrolepiota procera* (parasol mushroom), respectively (Biswas 1988).

CARBON DIOXIDE

It is commonly said that mushroom needs no fresh air during spawn growth. That is not true because accumulation of CO_2 only up to 4 per cent appears to be beneficial for the mycelial growth of *Agaricus bisporus* (Edwards 1978). The concentration of CO_2 must be under 0.08 per cent for fruit body initiation of button mushroom and that even less (0.05 per cent) in certain strains. Low CO_2 (not beyond 0.15 per cent) has been found as critical factor for obtaining good yield (Sohi 1988). Further, it is evident that 0.3–0.5 per cent CO_2 in the air above the casing soil results elongation of stipes and inhibition of sporophore initiation. Good aeration to minimize CO_2 concentration is also advocated for *C. indica* fruit body formation (Purkayastha and Chandra 1985). However, on the contrary, mycelial growth of *Pleurotus* sp. is stimulated by high CO_2 concentrations in air (Schanel 1970, Zadrazil 1975). It has been observed that CO_2 concentration in air up to 28 per cent (of volume) stimulates the growth of *P. ostreatus* and *P. florida,* while, that up to about 22 per cent stimulates the growth of *P. eryngie.*

IMPACT OF CLIMATIC FACTORS

Two distinguishing phases in the growth and development of cultivated mushrooms are the vegetative phase and the reproductive phase. During the vegetative phase enzymes secreted from the mycelium break down the lignocellulosic components into simpler soluble organic compounds, which are absorbed by the hyphae and used for metabolic reactions of the fungus for establishment. When the mycelia are established, they become ready to pass from vegetative phase to reproductive phase, and to provide strong physical support to the fruit body. Certain climatic factors trigger this change from the vegetative stage to reproductive stage. Amongst the factors, temperature, light and changes in atmospheric gases are significant in the transition from the mycelial state to reproductive (fruit body) state (Chang and Miles 1997). Higher carbon dioxide concentration and temperature are required for the vegetative growth of mushrooms while higher oxygen and more light are essential for fruit body initiation.

CHAPTER 4

Importance of Mushroom Cultivation

Mushrooms are considered as alternate food to combat the food shortage crisis of the increasing population in the world during the present era. These are available either by collecting them from their wild habitats or cultivating under controlled condition. In the former mode of availability, there exist certain risks in identifying the proper edible species. Since, several mushrooms growing in the same habitat are poisonous and these are difficult to distinguish without any expertness in this field. In certain cases, some mishaps are also happened by eating poisonous mushrooms. On the contrary, using the cultivated mushrooms as food has no such hazardous risks. All the cultivated mushrooms had been well studied for their edibility before commercialization. These mushrooms are cultivated with the known and correct edible species under controlled condition. In general, mushrooms are consumed as food for their excellent texture and flavour, and there is little awareness about the nutritional and medicinal attributes. However, the mushrooms provide us high amount of qualitative nutrition that is required for our growth and different medicines to combat diseases and to increase immune systems in our body.

NUTRITIONAL VALUE

Mushrooms have been recognized as delicious food of good quality protein. These are rich in vitamins, particularly in Vitamin C and Vitamin B-complex, and minerals. Further, these are the foods of low calorie with little fat, where sugar content is very less, starch and cholesterol are absent and ergosterol is present.

Protein

Protein is the most critical component contributing to the nutritional value of food (Crisan and Sands 1978). Its content in mushrooms mostly varies between 18 per cent and 30 per cent on dry weight basis (Table 4.1). About 80 per cent of the mushroom proteins, with certain exceptions, are digestible. In oyster mushroom (*Pleurotus sajor-caju*), 84.1 per cent of the protein is digestible (Thayumanavan and Manickam 1980). It is now established that the quality protein is nutritionally more important than its quantity. Mushroom protein comprises with most of the essential amino acids. These are rich in lysine and tryptophan, but deficient in sulphur containing amino acids, like methionine and cystine. The Essential Amino Acid index (EAA index) 98, and the amino acid score 89, of the mushrooms containing high amount of proteins are ranked just below the animal and milk proteins (Crisan and Sands 1978). The mushrooms such as, *Agaricus bisporus* and *Volvariella volvacea* contain higher proteins. Further, the nutritional index, which is calculated with the EAA index and protein per cent of mushroom protein is ranked just below the proteins of animal and soybean, while the biological values of mushroom proteins, which are calculated based on the reference proteins and EAA index, are considered as intermediate between vegetables and animal proteins (Crisan and Sands 1978, Rai 1995).

Amino Acids

Apart from the amino acids that combined in protein, mushroom fruit body also contains a high amount of soluble amino acids. Approximately 25–35 per cent of the total amino acids present in mushroom occur as free. These free amino acids are consisting of non-essential amino acids, amides and all the essential amino acids that required for our growth, but we are unable to produce them in our body. About 25–40 per cent of the amino acids are composed of essential amino acids (Crisan and Sands 1978).

Carbohydrate

Carbohydrate is the major fraction of the dry matter of mushrooms. It remains in two states: one is as nitrogen-free state and another is with nitrogen. The carbohydrate with nitrogen is mostly due to the presence of chitin in fungal cell wall. Thus, the nitrogen free carbohydrate is the only group of carbohydrates having importance in respect of nutritional point of view. Its content in cultivated mushrooms mostly ranged between 39.5 and 93.4 per cent on dry weight basis. In *Volvariella volvacea* the nitrogen free carbohydrate is 39.5 per cent while that in *Tremella fuciformis* is 93.4 per cent, in *Agaricus bisporus* is about 50 per cent and in *Pleurotus* sp. somewhere between 50 and 60 per cent (differently 46.6 to 81.8 per cent, Bano and Rajarathnam 1982). The carbohydrate of mushrooms is of less importance so far calories are concerned. However, in *Agaricus bisporus*, it is composed of a large variety of compounds, like pentoses, methyl pentoses, hexoses, disaccharides, amino sugars, sugar alcohols, sugar acids, some uronides and methyl sugars (Crisan and Sands 1978). Mannitol (a sugar alcohol) in mushroom remains very high as 9–13 per cent, while, free sugar is very less (0.5 per cent) and starch is absent. Although a wide variation presents in carbohydrate components of mushroom, it is mainly known for the presence of mannitol, glycogen and hemicellulose together with a small amount of reducing sugars (Mc Connel and Esselen 1947).

Fibre

Fibre has no food value but it acts as roughage in our food. Its content in mushroom is very high. In *Pleurotus* sp. it is ranged between 7.9 and 12.5 per cent (differently 7.5–27.6 per cent) on dry weight basis (Bano and Rajarathnam 1982, Rai et al. 1988). In *Agaricus bisporus* it is 9.1 per cent, while in *Volvariella volvacea* 11.9 per cent. The least fibre content (1.4 per cent) in mushroom is found with *Tremella fuciformis*.

Fat

Mushrooms possess a wide range of fat contents. The fat in mushroom may remain as low as 1 per cent in *Pleurotus eous* or as high as 8 per cent in *Lentinula edodes* (Bano et al. 1987, Crisan and Sands 1978). In *Agaricus bisporus* and *Volvariella volvacea*, the fat is about 3.0 and 6.4 per cent, respectively. The crude fat constitutes with all classes of lipids, such as free fatty acids, sterols, lecithin, sterol esters, phospholipids, mono-, di- and triglycerides. The fatty acids, like palmitic, stearic, oleic and linoleic acids are most common in *Agaricus bisporus*. Linoleic acid, which is a polyunsaturated fatty acid, is nutritionally more desirable. It is predominant with 70 per cent of fatty acids of neutral lipid fraction and 90 per cent of that of polar lipid fraction (Holtz and Schisler 1971). In *Volvariella volvacea,* linoleic acid is about 69.9 per cent of total fatty acids, while in *P. sajor-caju,* 62.9 per cent and *L. edodes,* 76.2 per cent (Huang et al. 1989). The total lipid contents in cultivated mushrooms vary between 1.6 and 3.1 per cent of dry weight of mushroom. Its content in *A. bisporus* and *V. volvacea* is higher than that in *L. edodes* and *P. sajor-caju* (Huang et al. 1985). In mushrooms, at least 70 per cent of total fatty acids are found as unsaturated and that is desirable in human diet. Mushrooms contain high amount of sterols. Among the sterols, ergosterol is most abundant and cholesterol is absent. The ergosterol contents in various mushroom species vary from 0.2–270 mg/100 g on dry weight basis (Crisan and Sands 1978). The commonly found ergosterols in mushrooms are provitamin D_2, provitamin D_4 and γ-ergosterol (Huang et al. 1985). These ergosterols are useful to produce vitamin D in human body in presence of Sun.

Vitamins

Mushrooms contain significant amount of vitamins, particularly thiamine (B_1), riboflavin (B_2), biotin, ascorbic acid and pantothenic acid. Vitamins A, D and E are generally absent in mushrooms, however, several species contain detectable amount of provitamin A, measured as β-carotene. Further, the ergosterols are present in mushrooms may convert into vitamin D in presence of ultra violet rays of Sun. In addition some species contain Vitamin K and B_{12} (Sumi 1933, Ramsbottom 1953, Crisan and Sands 1978).

Minerals

The high ash contents in mushrooms indicate that these are the rich sources of mineral contents. The minerals, like potassium, phosphorus and sodium are predominant, while calcium and iron are present in less quantity. Potassium alone constitutes 45 per cent of the total ash content of mushroom and this together with its followers, like phosphorus, sodium, magnesium and calcium, constitutes 56–70 per cent of total ash content (Chang and Miles 1989). In *Pleurotus* sp. potassium and phosphorus are the main constituents of ash (Bano and Rajarathnam 1982). In several mushrooms including *L. edodes* the calcium content is very poor. Iron content in certain mushrooms is to some extent more, but only one third of its total iron is in the available form (Anderson and Fellers 1942). Occurrences of several useful trace elements like, copper, zinc and manganese, as well as certain harmful heavy metals, like cadmium and lead have also been found in traces (Leh 1975, Bano et al. 1981, Bisaria et al. 1987).

MEDICINAL VALUE

Mushrooms have traditionally been used in China and Japan for the medicinal and tonic properties. Several cosmetic products and tonic beverages have also been produced in China from *Ganoderma* mushrooms. Pharmaceuticals worth $700 million are produced annually in Japan alone from the mushrooms like *Lentinus*, *Coriolus, Schizophyllum* and *Ganoderma*. Anti-tumour effects of *Lentinus edodes, Flammulina velutipes*, *Pleurotus ostreatus, Pholiota nameko, Tricholoma matsutake* and *Auricularia auricula* have been reported in their extracts (Cochran 1978). Recently, maitake (*Grifola frondosa*) and shiitake (*L. edodes*) mushrooms are reported as inhibitory to HIV virus of AIDS disease in USA and Japan. The water soluble polysaccharide of high molecular weight chemical, lentinan, present in the shiitake mushroom *Lentinus edodes,* reduces plasma cholesterol in human and controls blood pressure (Doshi and Sharma 1995, Wasser and Weis 1999). The active hypolipidimic principle in *L. edodes* is eritadenine [2(R), 3(R)-dihydroxy-4-(9-adenyl) butyric acid] affecting to lower cholesterol, triglyceride and phospholipid levels (Kaneda and Tokuda 1966, Suzuki and Oshima 1976). The mushroom has anti-cancer, anti-HIV, anti-diabetic properties and ability of boosting immune system (Chihara 1992, Jong and Brimingham 1993, Mizuno 1999). The anti-platelet substance, adenosine, of ear mushroom *Auricularia polytricha,* inhibits the platelet aggregation and is suggested for lowering atherosclerosis by eating this mushroom regularly (Markhija and Bailley 1981). The silver ear mushroom (*Tremella fuciformis*) has significant medical values in curing several diseases, like tuberculosis, blood hypertension, common cold, etc. It also gives nutrition required for skin care and improves personal appearance. Further, it increases the vigour and extends life span of human beings (Lee 1596, Chen and Hou 1978). The winter mushroom (*Flammulina velutipes*) is particularly known for its taste and medicinal properties. It cures liver diseases and gastro enteric ulcers. Further, it contains immunomodulatory, antitumor, tumor inhibiting and antibiotic substances (Yoshioka et al. 1973, Ying et al. 1987, Sharma et al. 2009). The compounds extracted from *A. bisporus, L. edodes, Coprinus comatus* and *Oudemansiella mucida* have antifungal and antibacterial properties. Mushroom as a low calorie-high protein diet with very little starch and sugars are the delight of the diabetic patient (Rai and Sohi 1988). Further, due to high potassium: sodium ratio, and low calories and fat content, mushrooms are preferred by the dietician for those suffering with obesity, hypertension and atherosclerosis (heart disease). Alkaline ash and high fibre contents in mushrooms also give relief from hyperacidity and constipation (Rai 1995).

Table 4.1 Proximate Compositions of the Important Cultivated Mushrooms*

Species	*Moisture*	*Dry matter*	*Protein*	*Carbohydrate*	*Fat*	*Fibre*	*Ash*	*Energy* (k cal)	*References*
Agaricus bisporus	90.1**	9.9	29.3	50	3	9.1	8.1	364	Rai and Sohi 1988
*Auricularia auricula**	16.4#		8.1	74.1	1.5	6.9	9.4	356	FAO 1970, Crisan and Sands 1978
Auricularia polytricha	90.5	9.5	4.4	68.0	4.4	11.9	6.0	384	FAO 1970, Crisan and Sands 1978
*Calocybe indica**	–	–	17.7	64.3	4.1	3.4	7.4	–	Doshi and Sharma 1995
Flammulina velutipes	89.2	10.8	17.6	69.4	1.9	3.7	7.4	378	FAO 1972, Crisan and Sands 1978
Lentinula edodes	90.0	10.0	17.5	59.5	8.0	8.0	7.0	387	Sawada 1965 Sugimori et al. 1971, Crisan and Sands 1978
Macrolepiota procera	84	16	20.4	62.0	3.6	7.0	6.9	373	Pilat 1954
P. eous	92.2	7.8	17.5	59.2	1	12	9.1	261	Bano et al. 1987
P. eryngii	93.2**	6.8	32.3	43.8	1.5	13.2	7.6	250	Rai et al. 1988
P. flabellatus	91.0	9.0	21.6	57.4	1.8	11.9	10.7	271	Bano et al. 1987
P. florida	91.5	8.5	18.9	58.0	1.7	11.5	10.7	265	Bano et al. 1987
P. membranaceous	89.6	10.4	20.2	58.1	1.4	12.5	7.8	269	Rai et al. 1988
P. ostreatus	92.9**	7.1	26.8	53.8	1.7	7.9	12.7	268	Rai et al. 1988
Pleurotus sajor-caju	90.2**	9.8	25.5	53.7	1.8	11.2	7.7	275	Rai et al. 1988
P. sapidus	91.6**	8.4	27.4	51.8	1.7	9.5	9.6	262	Rai et al. 1988
Tremella fuciformis	19.7#		4.6	93.4	0.2	1.4	0.2	412	FAO 1972, Crisan and Sands 1978
Volvariella volvacea	88.4*	11.6	30.1	39.0	6.4	11.9	12.61	374	Lee and Chang 1975, Crisan and Sands 1978

* Figures are in percentage on dry weight basis,
** Converted to dry weight basis,
# Moisture content of the dried and test samples.

CHAPTER

5

Mushroom Laboratory and Spawn Preparation

LABORATORY REQUIREMENTS

Cultivation of mushroom requires a well set up laboratory if one wish to be self dependent in this aspect. A clean laboratory with at least three rooms is required in maintenance of culture and spawn preparations. In case of large scale spawn production unit the build area 80′ × 50′ × 12′(h) is required. This area is divided into five separate small rooms for different activities, like working room, sterilization room, store room, culture room and incubation room. This space is sufficient to handle thousands of spawn bottles every day. However, in general a working laboratory with separate culture and incubation rooms are enough to cultivate mushroom. The working room should have been fitted with basin, sink, running water connection and electric wiring. This should have enough space for accommodating working table with chemical racks, microscope, low height table, pH meter, hot air oven, autoclave, BOD incubator, refrigerator, balance, almirah, chairs, etc. Certain instruments are used occasionally, e.g. microscope which is not required in maintenance of culture and preparation of spawn. However, if someone wants microscopic observation of disease causing organisms or weed fungi, then it is an important tool for identification. Here the most essential things which required in establishing a mushroom laboratory in private or government sectors are listed.

Working room

One working room of size 25′ × 16′ × 13′ (h) is enough to maintain culture and prepare spawn. The room is to be connected by electric line and water connection. The following equipment and facilities are required to be present in the room.

Working table: The working table is required for keeping balance, measuring chemicals, preparation of media, drying, mixing and filling in bottle/polypropylene packets of spawn materials and various other jobs time to time as required. Two such tables—one of low height for working under sitting condition and the other of high height for working under standing condition are required.

Autoclave: Autoclave is essential for sterilizing culture media and spawn medium. In certain cases, substrates are required to be sterilized (Figure 5.1).

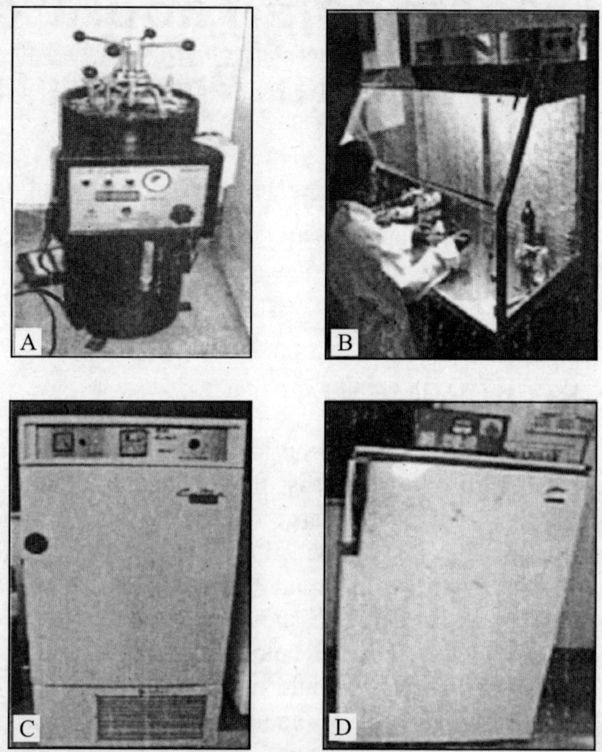

Figure 5.1 Mushroom laboratory equipment. **A.** Autoclave, **B.** Laminar-air-flow, **C.** BOD incubator, **D.** Refrigerator.

Balance: Measurement of various materials is done by balance. At least two different types of balance are required. One for measuring small amount material, such as electric or physical balance is required to measure chemicals as required in media preparation. Another type of balance, such as, pan balance, is required for measuring large amount of materials, like rice grain, wheat grain, etc. in spawn preparation. In case, spawn is to be produced on commercial scale, one larger balance (platform balance) is also required for measuring huge amount of spawn materials.

BOD incubator: This is a very vital equipment to grow spawn and culture at required temperature. This equipment is effective to maintain different temperature and the conditions required for the growth of mushroom.

Hot-air-oven: This instrument is required for drying different materials, sterilization of glass-wares, etc.

Gas oven: Cooking gas oven is required to prepare medium, boil spawn grains, water, etc.

Refrigerator: This is required to store mother culture for a little longer period.

Pressure cooker: Pressure cooker of 20 *l* capacity or more is required for cooking grains during spawn preparation.

Polypropylene bags of different sizes: Polypropylene bags of different sizes are required for spawn preparation, mushroom substrate preparation, etc. These bags are autoclavable and the spawn medium is filled in these bags. These are closed with an iron or hard autoclavable plastic rings and cotton plugs and sterilized in autoclave. Similarly for cultivating shiitake mushroom (*Lentinula edodes*) and winter mushroom (*Flammulina velutipes*) substrates are generally autoclaved after filling in polypropylene bags. The bags of 20 inch × 16 inch size are mostly used for filling spawned substrates during oyster mushroom cultivation.

Basket: Basket made of bamboo cane, plastic or iron net is required to drain out excess water from the boiled grains during spawn preparation.

Glasswares and other accessories: The mostly used glasswares are flasks of different sizes, measuring cylinders, culture tubes, petriplates, inoculating needle, culture tube stands, funnel, strainer, wire net basket, pH paper, non-absorbent and absorbent cotton, beaker, brown paper, spirit lamp, rubber bands and cotton thread.

Chemicals: The chemicals, like d-Glucose, agar-agar, malt extract, potassium hydroxide, hydrochloric acid, rectified spirit, methylated spirit, sodium hypochloride or mercuric chloride, calcium carbonate, calcium sulphate, magnesium sulphate, potassium dihydrogen phosphate, salicylic acid and water are essential.

Organic products: Potato, paddy grain and rice bran.

Medicine: Streptomycin sulphate and chloramphenicol.

Culture room: A closed room with UV germicide lamp and air conditioner. This room is separated from working room by partition.

Laminar-Airflow: This equipment where inoculation of mushroom culture and spawn medium is done under aseptic condition.

Incubation room: Beside a culture room, a room may be separated to incubate spawn and culture under room temperature. A few steel racks may be used to keep spawn packets. In large scale production of spawn, this facility is very much convenient. Further, certain mushrooms, like milky mushroom and paddy straw mushroom, required room temperature for their growth. A germicide UV lamp may be fitted inside to disinfect the room.

MUSHROOM CULTURE

Media

Before preparation of mushroom culture, medium preparation is essential where the vegetative growth of mushroom is being occurred *in vitro* at first. Most commonly malt extract agar is used for this purpose. However, Potato Dextrose Agar (PDA) medium is used in our case and gave more or less equal result. Other media which are mentioned below are also used for mushroom cultivation in certain cases as mentioned in the text with the respective mushroom.

PDA medium

Composition:

Peeled potato pieces:	20 g
Dextrose:	2 g
Agar (powder):	1.5 g
Tap water:	100 ml

Small pieces of peeled potato (20 g) are boiled in 50 ml water in a 250 ml conical flask for 10–15 minutes. Then, are strained by a strainer and the decoction is kept in a beaker. Agar powder (1.5 g) is boiled in rest of the 50 ml water in a conical flask with continuous stirring with a glass rod for 2–3 minutes. Now, both agar solution and potato decoction are mixed together. The previously measured dextrose (2 g) is also added to the mixed solution with stirring and made them homogenous. The volume up to 100 ml is adjusted with tap water in a measuring cylinder. The pH of so prepared medium does not require to be adjusted. However, if needed that can be adjusted to 5.5–6.0 with the addition of a few drops of 1/10N HCl or 1/10N KOH solution to make lower and higher pH by repeated checking with pH paper. The medium during hot condition is poured in culture tubes up to 1/3rd height keeping the culture tubes in culture tube rack for preparing slant and up to 2/3rd height of that for preparation of stab. The culture tubes are then plugged with non-absorbent cotton plug and wrapped 5–6 tubes with brown paper/two folds of used newspaper and rubber band. The wrapped tubes are then kept in a wire net basket and again wrapped the whole set as described except instead of rubber band cotton thread is used to tie. The medium is sterilized in autoclave at 15 p.s.i. for 15 minutes. A few clean petriplate pairs are also wrapped with brown paper and sterilized along with medium in autoclave.

Malt extract agar medium

Composition:

Malt extract:	2 g
Agar (powder):	1.5 g
Tap water:	100 ml

This is more simple method to prepare. All the ingredients are put together in a conical flask and boiled for 2–5 minutes. The volume is adjusted as earlier. The rest of the procedure is same as described with PDA medium.

Malt yeast extract agar medium

Composition:

Malt extract:	1.5 g
Yeast extract:	0.3 g
Agar:	2.0 g
Tap water:	100 ml

This is more simple method to prepare. All the ingredients are put together in a conical flask and boiled for 2–5 minutes. The volume is adjusted as earlier. The rest of the procedure is same as described with PDA medium (Chen and Hou 1978).

Corn meal agar medium

Composition:

Corn meal:	10 g
Dextrose:	1.0 g
Agar:	2.0 g
Tap water:	100 ml

Corn meal (10 g) are put into 100 ml water and kept at 60°C for two hours. The extract is filtered with a muslin cloth and the other ingredients, like dextrose (1.0 g) and agar (2 g) are added to the filtrate. Now, the filtrate mixture is boiled for 2–5 minutes. The volume is adjusted as earlier. The rest of the procedure is same as described with PDA medium (Chen and Hou 1978).

Rice meal agar medium

Composition:

Rice meal:	10 g
Dextrose:	1.0 g
Agar:	2.0 g
Tap water:	100 ml

Rice meal (10 g) are put into 100 ml water and kept at 60°C for two hours. The extract is filtered with a muslin cloth and the other ingredients, like dextrose (1.0 g) and agar (2 g), are added to the filtrate. Now, the filtrate mixture is boiled for 2–5 minutes. The volume is adjusted as earlier. The rest of the procedure is same as described with PDA medium (Chen and Hou 1978).

Slant Preparation

After sterilization, a pinch of streptomycin sulphate is added aseptically in each tube medium with the aid of inoculating needle, spirit lamp, rectified spirit, absorbent cotton and laminar-airflow. The tubes then rubbed for a while between hand palms and placed on a one inch high wooden or thermocol strip in slanting position for solidification on laminar-airflow-desk prior illuminated with germicide UV lamp for an hour.

Pouring of Stab Medium in Petriplate Pairs

The sterilized stab medium is poured on lower half of petriplate after adding a pinch of streptomycin sulphate on that with the help inoculating needle. The medium after pouring in petriplate is stirred for well mixing with streptomycin sulphate by gently moving on the laminar flow desk. Now, the whole set is allowed overnight for solidification and cooling on laminar flow desk.

Mushroom Isolation

Isolation is done from naturally growing fresh mushrooms in any of the following two methods:

Pileus tissue culture

Fresh fruit body may be surface sterilized with 0.1 per cent $HgCl_2$ or 0.5 per cent Sodium hypochloride. However, in our case rectified spirit in a cotton lump is used for surface sterilization by gently rubbing over the surfaces. Now, the pileus is teared into two halves with hands and the inner most tissue of the pileus is taken out by red hot burn inoculating needle or by pointed burned forceps. The tissue is placed on the solidified medium in slant, aseptically. The inoculated slants are labelled, wrapped with brown paper again and placed in culture shelf at room temperature or in BOD incubator at suitable temperature, depending upon the nature of mushroom species (Figure 5.2).

Figure 5.2 Spawn and culture on racks. **A.** Culture, **B.** Planting spawn.

Spore culture

Spore culture can be raised from single spore or multiple spores. For raising mono or multispore cultures, spores are collected from a large and healthy mushroom. The mushroom after surface disinfection with the help of rectified spirit in cotton lump is mounted over a sterile petriplate directly or over a sterile dry filter paper in sterile petriplate for spore deposition called **spore print**. Now, with the help of inoculating needle and sterile razor blade a small piece of filter paper with spore mass or spore mass scratching from spore print on petriplate is transferred to PDA medium in slant or at the centre at petriplate in medium, aseptically, a few days (nearly 10 days) after mycelia grow out at the point of inoculation.

In case of single spore culture, a small piece of filter paper with spore mass or spore mass collecting from spore print in petriplate are immersed in 10 ml of sterile distilled water in culture tube and shook that well for separation spores from each other. Now, by serial dilution method the spore density is minimized. One ml of diluted spore suspension is poured in each sterilized petriplate pairs. Then, the stab medium (agar 2 per cent, 6 ml or equal amount of Lambert's medium prepared by the ingredients of 10.0 g glucose, 0.5 g $MgSO_4.7H_2O$,

1.9 g KH_2PO_4, 20 g agar and 11 ml distilled water) cooled, but not solidifying by hand palm rubbing is poured in the petriplate pairs with spore suspension. Now, by rotating petriplates on working desk in laminar-airflow the spore suspension and the medium are mixed well. They are allowed to solidify and the spores to germinate at desired temperature. The spores which remain on the surface of medium germinate and mycelia become visible after 10 days of incubation. In case of contamination, if any, a filter paper piece larger than the size of contaminated fungal or bacterial colony is soaked in 10 per cent salicylic acid in alcohol and compressed on the colony by a forcep. A small bit of mycelia germinated from single spore is cut under microscope and transferred to PDA slant with inoculating needle. However, it is to mention that the single spore culture of heterothallic mushrooms will not develop into fruit bodies, although, it is very useful for strain improvement of that mushrooms by developing dikaryotic mycelia in hybridization programme.

Purification of Culture

The mushroom culture so prepared by different methods might be contaminated with bacteria. So, purification is needed. This is done by subculturing on 2 per cent agar medium in petriplate. The slow growth of bacteria (if any) allows them to remain confined at the centre. After 14 days mycelia from advancing zone are cut with 5 mm cork borer and transferred to PDA or malt extract medium in slant.

CONSERVATION OF MUSHROOM CULTURES

Successful mushroom production depends upon the proper maintenance of pure culture that is capable of producing fruit bodies of high productivity (Jong 1978). There are four main ways by which the cultures are conserved in laboratory.

Conventional Methods

Two conventional methods for conserving the mushroom cultures are practiced in most of the laboratories all over the world.

Periodic transfer

Stock cultures are maintained in an actively growing state under optimum laboratory conditions on suitable solid medium (PDA or malt extract agar medium) using periodic transfer of mycelia at reasonable time intervals.

Isolation

Pure cultures are reestablished under field cultivation by using spore or pileus tissue culture from the freshly harvested sporophore.

However, these conventional methods are laborious, costly and risky. Partial loss in productivity and vigour may occur due to degeneration and mutation during prolonged vegetative propagation of stock cultures or due to genetic recombination and selection in continuous field cultivation of reestablished cultures (Burnett 1975).

Preservation in Paraffin Oil

In this method, the actively growing mycelia are preserved under sterilized liquid paraffin (paraffin oil) of specific gravity 0.83–0.85. The paraffin oil is sterilized twice in two consecutive days in an autoclave at 15 lbs p.s.i. for 15 minutes each. The mushroom mycelia of 14 days in slant are considered for preservation. The sterilized paraffin oil is poured aseptically inside the culture tube grown with mushroom mycelia as such that the paraffin oil remains about 1 cm above the medium. Alternatively, in certain cases, 0.5 cm mycelial discs are cut from myceliated medium in petriplates and suspended in 4–5 ml of sterilized liquid paraffin in a small sterilized vial. The paraffin oil blocks the exchange of oxygen between the mycelial surface and the atmosphere in the container, thus retards metabolism. This also prevents desiccation of the agar medium. The paraffin oil-filled mushroom culture is kept at room temperature (28 ± 2°C) for storing about 1 year or in refrigerator at 5°C for storing about 4–5 years.

Freeze-drying and Freezing Method

In this method Lyophilizer, semi automatic sealer and liquid nitrogen refrigerator are the major equipment to be used. It is very economical for long-term preservation. The spores collected from fresh mushroom fruit bodies are frozen and simultaneously dried under low pressure in a vacuum atmosphere to remove water from spores. Under this condition spores remain dormant and viable for longer period.

In case of non-sporulating fungi, mycelial plug from advancing zone is cut with 5 mm diameter cork borer and transferred to thick walled borosilicate glass ampoules for freeze drying and then it is sealed with semiautomatic sealer to store in liquid nitrogen.

Cryogenic Freezing Method

Cryogenic freezing method is done to avoid freezing injury in freeze-drying method. In this method culture in agar slants is scraped in 10 per cent glycerol or 5 per cent dimethyl sulfoxide solution in distilled water to obtain mycelial suspension. Now, the suspension of mushroom cultures are filled in thick borosilicate glass ampoules and precooled to 5°C for 30 minutes before sealing with semi automatic sealer. The sealed ampoules are kept in liquid nitrogen and freezed slowly @ 1°C/minute up to –35°C, then rapidly up to below –100°C. After the ampoules have been frozen, they are immediately transferred to storage in liquid nitrogen at –196°C.

Preparation of Spawn

Spawn is the vegetative mycelium grown on a suitable medium. Its preparation requires a well equipped laboratory a little bit away from mushroom growing unit. In case of large scale production of spawn, a built up areas of about 80′(l) × 50′(b) × 12′(h) is required. This area is required to be subdivided into different work areas such as inoculation room, incubation room, autoclaving and boiling room, filling room, cold room, etc. Generally spawn preparation is done in two steps, mother spawn and planting spawn, with the use of same medium or different media. Various kinds of spawn media are successfully used in different laboratories.

Here, spawn prepared with paddy grain medium is described due to its suitability in respect of its easy availability, low price in all states and efficiency to give equally good production of fruit bodies as with wheat grain based medium (Biswas and Singh 2008).

Mother Spawn

Two kg fresh paddy grains are boiled in double volume of water for 25–30 minutes in a pressure cooker. After boiling, the paddy grains are placed in a small bamboo made basket keeping that on a sink for over night to drain out the excess water and to dry the surface area to minimize stickiness. Sixty gram of calcium sulphate (gypsum), twenty gram of calcium carbonate and 500 mg chloramphenicol are mixed separately in a glass bowl and added to the boiled paddy grain on a clean platform with thorough mixing. The mixed substrates are the spawn medium and filled up in 500 ml milk bottles or conical flasks @ 250 g/bottle. These bottles are plugged with non-absorbent cotton plug and the mouth is wrapped with a piece of brown paper and a rubber band. The bottles are now sterilized in an autoclave at 15 p.s.i. for 2 hrs. After sterilization on the next day, the spawn medium is inoculated, aseptically, with a bit of mycelial colony, taking out from mushroom culture grown in slant or in petriplate medium. The inoculated spawn bottles are incubated for 10 days at a suitable optimum temperature in BOD incubator. They are shook twice beating them on hand palm on 3rd and 7th days for even growth of mycelia and to observe contamination if any.

Planting Spawn

In this type of spawn, substrate may remain same as in mother spawn or it may be prepared by mixing the boiled grains with half dose of calcium carbonate (5 per cent) and calcium sulphate (15 per cent). However, unlike mother spawn, addition of chloramphenicol is not needed in any kind of the substrate mixture. The spawn medium is filled up in polypropylene packets @ 150 g/packet. The packets are plugged with non-absorbent cotton plug supported by an autoclavable plastic or metallic ring. After sterilization as earlier, inoculation is done by about 10 g of 10-day old mother spawn. Very old mother spawn is not suitable due to hardness in breaking that with a glass rod. The incubation is done at room temperature in culture room or at 25 ± 1°C in BOD incubator, as suitable. The spawn becomes ready after 14–21 days of growth.

Note: *The methods discussed in this chapter are based on paddy grain medium, as followed by the authors. Apart from paddy grain, several other grains like wheat, jowar, and bajra are most frequently used for spawn preparation. In commercial scale wheat grains are most preferred. Use of antibiotics (chloramphenicol) as described with mother spawn preparation is not generally practiced. Simply the boiled grains (boiling for half an hour) after draining out the excess water and drying the grain surface in air are mixed with* 10 g *calcium carbonate* ($CaCO_3$) *and* 30 g *calcium sulphate* ($CaSO_4$) per kg *of dry substrate. Further, in case of planting spawn preparation the amount of calcium carbonate and calcium sulphate is minimized. In that case,* 1 kg *base substrate is mixed with* 5 g $CaCO_3$ *and* 15 g $CaSO_4$.

CHAPTER 6

Button Mushroom Cultivation

Button mushroom is also known as common mushroom or temperate mushroom, it is the fungal species, *Agaricus bisporus* (Lange) Sing. It is the most popular mushroom in the world. It has high demand in the market worldwide due to its unique shape and feature, like white creamy and large button shaped soft fruit body with good flavour, taste and nutritive values. As per the production it ranked first in the world. During the year 1997 its world-wide annual production was 1,956,000 metric ton (mt), sharing 31.8 per cent of the total production of mushrooms (Chang 1999). It grows in temperate areas of the world. In addition, one more similar species, viz. *Agaricus bitorquis* (Quel.) Sacc., with little harder silky white sporophore and commonly known as street mushroom or pavement mushroom, is also cultivated in the name of temperature tolerant white button mushroom in tropical and subtropical areas where temperature is to some extent higher. The temperature requirement of the latter one is 30°C for optimum vegetative growth and 25°C for developing sporophore.

The cultivation of button mushroom (*A. bisporus*) first originated in Paris (France) around 1650, where melon growers observed spontaneous appearance of this mushroom on used compost of melon crops (Delmas 1978). Later, about one hundred and thirty years after it was observed that the mushroom could grow in dark without any light and the cave cultivation was started. Now, the button mushrooms are cultivated both in cave and house on decomposed pasteurized compost made from various agricultural residues enriched by nutritive ingredients. In India the button mushroom cultivation is mainly confined to the hills of all Himachal Pradesh, Jammu and Kashmir, Haryana and Uttar Pradesh. In the current treatise, *A. bisporus* is categorized as 'white button mushroom' while *A. bitorquis* is referred as 'temperature tolerant white button mushroom' in order to distinguish the two types of button mushrooms.

MORPHOLOGICAL AND OTHER CHARACTERS

White Button/Common Mushroom (*Agaricus bisporus*)

The mushroom generally grows in meadow, pastures and in open fields during September or October. It grows solitary in ring, but sometimes two or more fruit bodies may appear from the same base. The pileus/cap is 5–10 cm broad, convex, nearly flat at centre on maturation, centrally stipitate, at first white, later turns into pale brown, naked to finely scaly, dry with thick projecting margin. Margin extends beyond the gills, fringed, sometimes fibrillose-striate or crenate. The flesh is firm, white slowly becoming salmon-pink, odour acidulous, taste pleasant. Gills are free, crowded, narrow to moderately broad, whitish turns to pinkish at young and, black at maturation. Veil present, well developed, soft, membranous, white to slightly brownish, usually superior, eventually inferior, peeling downwards, more or less distinctly double. Stipe 3–5 cm long, 10–15 cm broad, broader near base, white, annulate, not volvate, sometimes flesh coloured at apex, sometimes floccose below annulus, pithy becoming fistulose, whitish. Annulus white, thick, narrow, striate above; spores 6.3–8.5 × 5.0–6.8 μ, short-ellipsoid, smooth, brown, chocolate brown or mummy brown in mass, usually two spores on a basidium. Basidium is with two sterigmata of which each bearing one basidiospore at tip. Pleurocystidia are absent. Cheilocystidia are broadly clavate, hyaline to brown, 17–44 × 8–12 μ (Smith 1978).

This mushroom has limited genetic variability in the cultivated strains all over the world. The strains are mostly developed by selecting isolates from different localities all over the world. They are of different colours: white, cream and brown. The strains with white coloured and smooth surfaced fruit bodies are predominant and mainly used for fresh consumption. However, the yield of this kind is to some extent lower than those with cream or brown coloured caps with scales. The latter ones are mostly used for preservation in canning. The hybrid strains which are used for cultivation are produced by mixing the spawns of different selected strains. The commercial strains, like S-11, S-130, S-140, S-649, S-791, CM-1, CM-5, CM-10 (hybrid), A-15 (Sylvan, hybrid), U-3, X-13, Delta, etc. are used in India. Amongst them, S-11, S-649 and S-791 are good strains in India (Kapoor 1989). Further, a single spore isolate (DK-35) of Strain S-11 is found better than the original strain (Suman 2005).

Temperature Tolerant White Button Mushroom (*Agaricus bitorquis*)

The mushroom grows solitary or in cluster in grassy places and road sides. Cap/pileus is 5–12 cm broad, at first hemispherical, then convex, and at last expanding to nearly plane with slightly depressed at centre, centrally stipitate, surface dry, silky fibrillose, white, occasionally areolate in age and disc grayish. The flesh is thick, hard, white, slowly becoming vinaceous near gills, colour unchanged when cut, odour and taste mild. Gills crowded, at first pinkish grey, then pale dull vinaceous pink and finally blackish brown, free, narrow to moderately broad. Stipe short, tapering towards base, solid, whitish, sometimes vinaceous brown, glabrous. Annulus is at median position of stipe, very thick, white, double and edge of second ring prominent. Basidium is four spored. Spores are subglobose to broadly ellipsoid, purple brown, blackish brown in mass, 5–6 × 4–5 μ in size. Pleurocystidia are thin walled, club shaped, sometimes saccate, scattered to rare, 28–36 × 9–14 μ in size. Cheilocystidia club shaped, 10–18 × 6–12 μ in size. Hyphae are without clamp connection.

The temperature tolerant button mushroom is resistant to viral diseases; it requires about 5ºC more temperature for fruit body formation than the temperate button mushroom. It can also grow in the condition with higher CO_2 concentrations. It is harder than the latter one, thus it is suitable to transport fresh in market. Heterothallic nature of its mycelia creates great advantage in breeding work to grew new strain by inter strain breeding, which is not easy with secondarily homothallic species, like *A. bisporus*. However, the *A. bitorquis* has some disadvantages. It requires prolonged growing period and gives lower yields. The important cultivated strains of subtropical mushrooms are Horst B-30, K-26, K-32, K-46, NCB-6 and NCB-13.

FACILITIES REQUIRED

Button mushroom cultivation following the improved method of technologies requires the facilities like composting yard, bulk chamber or peak heating room, casing pasteurization chamber and environmentally controlled growing room. However, in case traditional method following long method of composting, only composting yard and seasonal growing room are required.

Composting Yard

Composting yard is a place where compost preparation is done. It is only a tin covered (GI roofing) shed on RCC pillars. It has no walls, but has a cemented floor which gently sloped towards a drain that ended to a common tank (guddy pit) to collect and reuse the liquid manures which leaked from the substrates during watering and mixing. The drain runs on two sides of platform to facilitate periodic cleaning. A composting yard of 100′(l) × 60′(b) × 20′(h) size is required to prepare compost in three simultaneous stacks of 8 tons each along with space for wetting straws and pile preparation. This space is also adequate to produce 25 metric tons of ready compost required for growing mushrooms in cropping room of 60′(l) × 30′(b) × 10′(h) size.

Pasteurization Room/Chamber/Tunnel

Pasteurization chamber/tunnel is required in phase-II for pasteurizing and conditioning the compost in short method of compost preparation. Two different systems of pasteurization units are common.

Peak heating room

In this system a simple insulated room has arranged for steam jets and air through circulating fan. An insulated room of size 24′ × 16′ × 8′ can accommodate 250 trays for peak heating (Singh 1983). The room is attached with a boiler outlet for injecting steam and the devices which specially made with ventilating fan to introduce and remove air by operating valves. An electric blower (½ hp) is attached with a perforated polythene pipe length-wise of the room. Alternatively, one or two circulating fans are used to re-circulate the air inside the room. The compost is filled in trays for pasteurization. In this system, all the operations like pasteurization, conditioning, spawn running, casing and fruiting are done in the same room. This system is suitable only for small scale of production.

Bulk pasteurization chamber/tunnel

This system is required for pasteurization of compost in mass. The tunnel size 45′ × 8′ × 13′(h) is mostly used for 25 tons of compost preparation. In this system the chamber/tunnel has two floors, the real floor and grated floor (Figures 6.1 and 6.2). It is a specially built tunnel, which is properly insulated and provided with steam connection and air blowing system. The real floor is made up of sand, broken stone or brick, concrete floor, glass wool, isolating membrane of PVC sheet one layer, concrete floor and finishing top, successively in layers. The outer wall of the tunnel is made up of bricks of 10″ thickness. The wall is separated from the concrete floor by 1 cm thick polystyrene battens which later burnt off and filled up the gap with bituminous sealant. The roof is made of 10 cm thick RCC. The inner surfaces of the roof, wall and real floor are covered with 5 cm thick insulating material, like asbestos

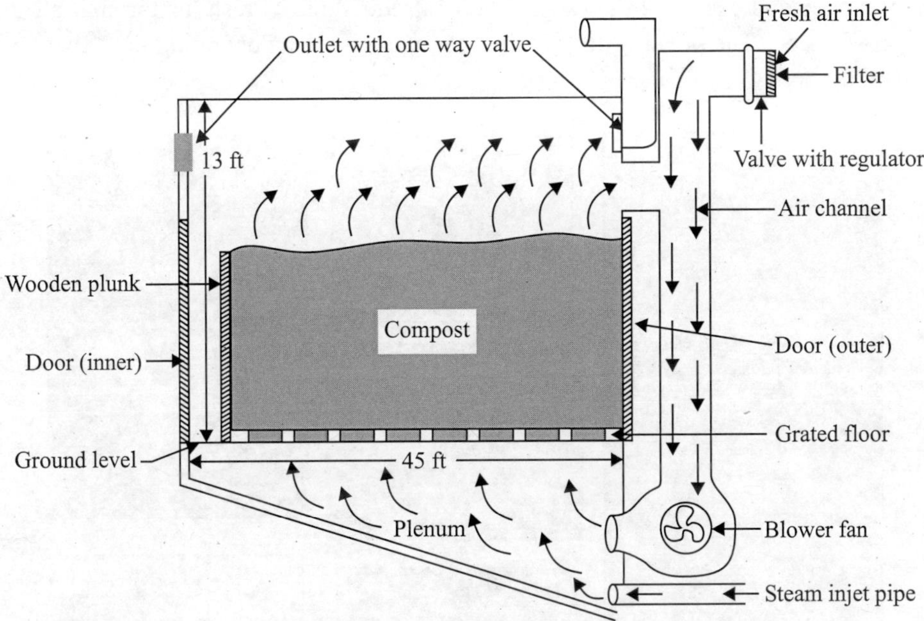

Figure 6.1 Diagram of bulk chamber.

sheet. The real floor of the chamber is with 2 per cent slope towards the front to facilitate draining out the water through the passage below air inlet. A wooden/iron grated floor with 20–25 per cent gaps is placed 50 cm above the real floor. The composts are kept on the grated floor for peak heating. The space between the grated floor and the real floor is used as plenum. The ceiling, inside wall surface and floors (both grated and original ones) are painted with two to three coatings of bituminous paints for well run-off the water which formed as droplets due to condensation of vapour. The tunnel may be one door system, while compost filling is done manually or double door system, if compost filling is done by conveyor belts. In case of double door system, there are two opposite doors on the short sides of the bulk chamber. The doors are made up of wood and are properly insulated with 5 cm asbestos sheet which again are covered by aluminum sheets. The doors are fixed with iron angles on the wall and can be closed hermetically. The height of the door on back wall is kept 8 ft. high to facilitate the filling with slated conveyor belt. Outside the chamber in bunker below the plenum a centrifugal blower fan with a speed 1450 rpm, which supported water pressure 125 mm and produced 150–200 m^3 air per hour, is fixed. This blower fan blows steam which is inserted in the plenum by steam injet pipe and fresh air drawing through an air channel. A filter is fitted at the mouth of the air channel to clean the air. The boiler, which produced steam, may be placed fixed beside the blower machine or in a separate boiler room. A boiler with production capacity 150 kg steam per hour is required for the aforementioned size of bulk chamber. The steam injet pipe is fitted below the blower. Two exhaust vents, one above other, are kept above the front door. The lower vent is opened in a channel for recirculation of steam and hot air by blower through plenum and the upper one with one way valve is opened in the air and used to exhaust gases on introduction of fresh air through air channel via dampers (filters). The opening and closing of vent is done by operating the valve manually from outside.

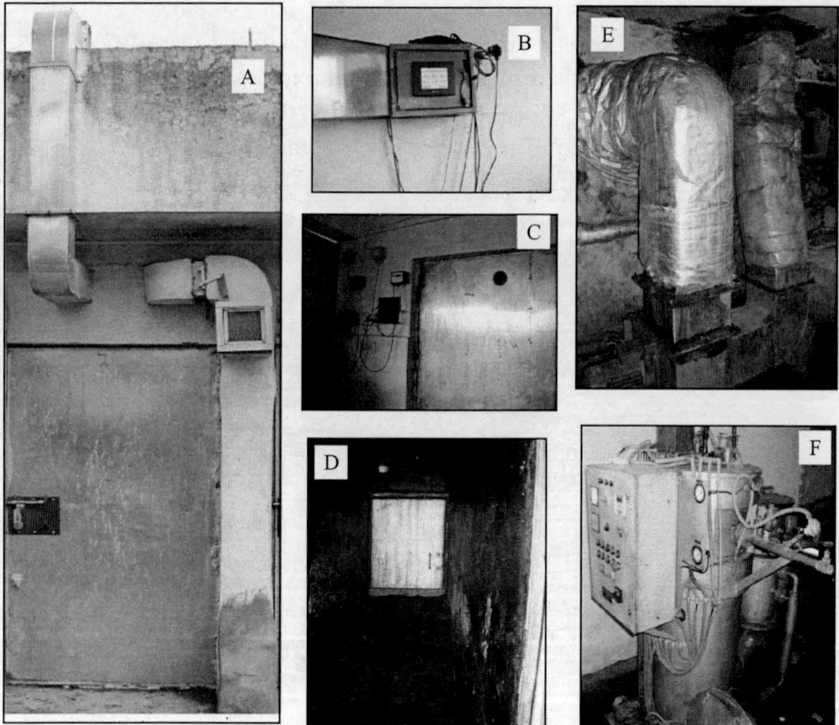

Figure 6.2 Bulk chamber. **A.** Front view, **B.** Temperature recorder, **C.** Back view, **D.** Inner view, **E.** Blower at bunker and **F.** Boiler.

Casing pasteurization chamber

This is just like bulk pasteurization chamber with insulating materials on the walls, door and floor, but of smaller size, like 15′ × 9′ × 13′(h). The plenum below the grated floor is kept 25 cm in depth. Besides the chamber a water tank with a tap and drainage pipe is needed to steep and wash the casing material. This washing is needed to avoid salt accumulation in the casing material. The washed and steeped casing materials are placed in trays on the grated floor above plenum. The steam and air recirculation devices are same as for bulk chamber.

Mushroom Growing Room

White button mushroom is cultivated in caves, in ordinary brick houses/sheds or in environmentally controlled growing rooms. As cave cultivation is not practiced in India so it is not discussed. In India most of the mushroom growers are seasonal and they grow mushrooms in ordinary brick-made or thatched houses. However, for better yield specifically modified mushroom growing rooms are required.

Seasonal mushroom growing room

Ordinary mushroom growing room is suited for seasonal cultivation of white button mushrooms. This type of mushroom growing room is a simple brick constructed building with cemented floor and RCC roof or asbestos shed with false ceiling. The suitable size of growing room is 25′ × 15′ × 10′(h). The room is to be well ventilated with the use of perforated vent pipe and exhaust fan, the latter one fixed in opposite direction to force air inside the room. Two vents covered with iron/nylon mesh are kept 2–3′ above the floor level on the back wall for releasing the excess air to outside. The shelf can be made of wood or steal and in three to four tiers system. Damper at different places inside the room is placed to raise humidity inside the room. This mushroom house is suited for seasonal cultivation of mushroom (Figure 6.3).

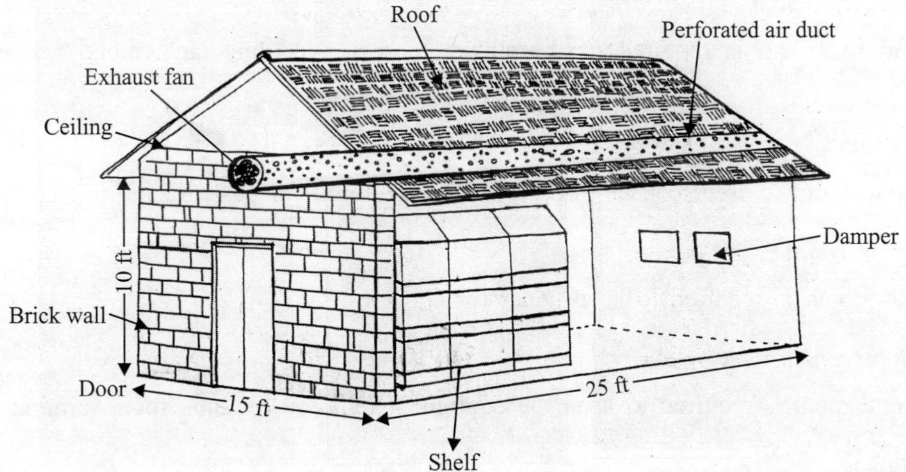

Figure 6.3 Diagram of growing room.

Round the year mushroom growing room

Environmentally controlled mushroom growing room is required for round the year production of button mushrooms. This type of mushroom growing room is prepared with proper insulation in floor, walls, ceiling and door, that can be closed hermitically with the use of rubber gasket. The room size of 60′ × 30′ × 10′ is required to grow mushrooms from 25 metric ton compost. The floor is prepared in the same way as described in bulk chamber preparation, but in this case underground drainage system is prepared lengthwise at the middle of the room by incorporating a broad polythene pipe before placing RCC layer and the floor surface is kept slanted towards the middle. The wall is made up of brick (10″) and covered with glass wool/polyurethane with the use of hot coal tar. Cement plastering of 2 cm is done on the top of this insulator to give a smooth finish. This is again covered with bituminous paint for providing barrier of vapour entry. The roof top is smeared with tar, followed by application of 10 cm loose soil, 5 cm thick mud capping which is again covered with tiles or GI sheets, respectively. The room is kept well ventilated by fixing perforated polythene vent pipe length wise above door height of the room. An exhaust fan (24″ diameter) operating by 1.5 H.P. motor is connected at the mouth of the vent pipe. Two vents covered with iron/nylon mesh and insulated valves are present 2–3′ above ground level on the back wall of the room in order to get release of inside air to outside. Cooling and heating devices are made respectively by installing cooler, whose cooling coils are immerged in chilled water; and heater, whose heating coils are connected to the steam pipe line of boiler. Humidifier inside the room is fitted for raising humidity. The room has no window. It is kept in illuminated condition by fixing fluorescent tubes vertically on the walls. In addition, a movable strong light is arranged to inspect beds.

Apart from the above three major facilities, several other miscellaneous facilities are also required in establishing button mushroom farm. For example,

Air condition room

Air condition room is required for keeping compressor, cooling tank, motor and electric panel.

Boiler room

A separate room is used to keep boiler which is required for pasteurization of compost.

Generator room

Generator room is required to keep generator and diesel.

Equipment/machinery room

A separate room is required to keep the equipment required for mushroom farming.

Store house

Store house is required to store different ingredients of compost preparation.

Water tank

A separate water tank is required for keeping fresh water and soaking of straw.

Packaging room

For packing of mushrooms a separate room is required.

Spawn production unit

A well set up spawn production unit is required to produce disease free, quality spawn.

Composting machineries, appliances and facilities

Certain facilities and equipment are required during composting. These are discussed as follows:

Floor grids and wooden panels: These are required to pile the compost in proper shape and size. The floor grids [measuring approximately 6(l)′ × 5′(b)] are made up of wooden slats. Three side panels, one is back panel measuring 5′ × 5′ and other two are side panels (6′ × 5′) are so prepared that these can remain as erecting position with the tilted wooden support on back. During compost preparation, these panels are placed in such a way that they make a dice with one side open and the tops of the panels remain slightly slanted towards centre. The wooden slates are used at the base to provide air inside the compost heap through the few centimetres gap as formed between the concrete floor and compost heap. Traditionally, the floor grid is not used in compost preparation. There, the wooden panels are placed on a cemented floor to make the dice as above.

Compost turner: This is required for piling the compost, mixing the ingredients and turning the compost after certain intervals.

Fork: This is used for mixing straw with water and turning the compost.

Water: Running water tap with polythene delivery pipe is essential to soak the substrates.

Container: This is required for carrying and heaping the ingredients during mixing and wetting.

Thermometers: Three to four numbers of long (60–80 cm) centigrade thermometers are required to measure the temperature in the compost pile.

COMPOST PREPARATION

There are mainly two different methods of compost preparation. First one is long method which does not require pasteurization, and the second one is short method where pasteurization is needed. These methods provide adequate levels of nutrients to support the growth of button mushroom (*Agaricus bisporus*). However, of the two methods, the second one is much effective to produce quality compost. In long method the mushroom gives low yield and even in some cases crop failure may occur. Various materials in different formulations are used for compost of white button mushroom. However, the ingredients which are required for compost preparation are broadly classified into three major groups as follows:

Ingredients

Base material

Straws of different agricultural products, like wheat, rice, oat, maize stalk, sugarcane bagasse, etc. are in this group. The straws are needed to cut into 5–8 cm long pieces in order to create compactness in compost heap.

Supplements

Different types of nutrients are added to the base material to enrich nutritive contents in the compost. The supplements are described as follows:

Manure: It is used to supply nitrogenous material. Chicken manure, horse manure, pig manure and sheep manure are used in compost preparation. Horse manure is the best material for compost preparation, but it is not available everywhere so chicken manure is used mostly. Chicken manure is most suited in short-term composting system, while cattle manure, on the other hand, is not suitable for compost preparation in any system of composting (Vijay and Gupta 1995).

Carbohydrate nutrients: Different types of materials, such as molasses, malt sprouts, potato wastes, etc. are used to supply carbohydrate nutrients in the compost.

Nitrogenous fertilizers: Nitrogenous fertilizers, like ammonium sulphate, calcium ammonium nitrate and urea are used to enhance microbial growth during composting.

Concentrate meals: Animal feeds, like rice bran, wheat bran, cotton seed meals, soybean cake, mustard cake, brewer's grain, are used as supplements to the base material. They are used to provide both nitrogenous and carbohydrate foods.

Rectifying materials: Different types of other materials are also required to rectify the compost. Gypsum and calcium carbonate are used to maintain compost pH and to devoid greasiness/stickiness of the compost. Insecticides and nematicides are used to control pest attacks.

Formulations

Different formulations of compost materials are given by different workers at different times and on the basis of ingredients they are categorized into two groups one is manure/natural compost where horse manure is used and other is straw compost/synthetic compost devoid of horse manure.

Formulas of manures

Formula 1: Horse manure compost (Van lier et al. 1994)

- Horse manure: 800 kg
- Chicken (broiler) manure: 80 kg
- Gypsum: 20 kg
- Water: 240 l

Formula 2: Horse manure compost (Beyer 1999)

- Horse manure: 80 ton (wet wt.)/50 ton (dry wt.)
- Poultry manure: 7.5 ton (wet wt.)/6.0 ton (dry wt.)
- Brewers grain: 2.5 ton
- Gypsum: 1.25 ton

Formula 3: Natural compost (IARI, in Vijay and Gupta 1995)

- Horse manure: 1000 kg
- Wheat straw: 350 kg
- Urea: 3 kg
- Gypsum: 30–40 kg

Formula 4: Natural compost (IARI, in Vijay and Gupta 1995)

- Horse manure: 1000 kg
- Wheat straw: 500 kg
- Chicken manure: 300 kg
- Urea: 7 kg
- Brewer's grain: 60 kg
- Gypsum: 30 kg

Formula 5: Natural compost (Hayes and Randle 1969)

- Horse manure: 1016 kg
- Chicken manure: 102 kg
- Molasses: 38 kg
- Cotton seed meal: 15 kg
- Gypsum: 15 kg

Formulas of synthetic composts

(i) For long method of composting

Formula 6: Synthetic compost (Agarwala 1973)

- Wheat straw: 300 kg
- Calcium ammonium nitrate: 9 kg
- Super phosphate: 9 kg
- Urea: 4 kg
- Sulphate of potash: 3 kg
- Wheat bran: 30 kg
- Gypsum: 30 kg

Formula 7: Synthetic compost (Seth and Sandilya 1975)

- Wheat straw: 300 kg
- Calcium ammonium nitrate: 6 kg
- Urea: 2.4 kg
- Super phosphate: 2.5–7.5 kg

- Sulphate of potash: 3 kg
- Brewer's grain: 15 kg
- Gypsum: 30 kg

Formula 8: Synthetic compost (Kachroo et al. 1979)

- Paddy straw: 300 kg
- Molasses: 12 kg
- Urea: 4.5 kg
- Wheat bran: 50 kg
- Cotton seed meal: 5 kg
- Muriate of potash: 2 kg
- Gypsum: 15 kg

Formula 9: Synthetic compost (Munjal and Anand 1979)

- Wheat straw: 300 kg
- Calcium ammonium nitrate or ammonium sulphate: 9 kg
- Urea: 3.6 kg
- Sulphate of potash or muriate of potash: 3 kg
- Super phosphate: 3 kg
- Wheat bran: 30 kg
- Spent brewer's grain: 40 kg
- Gypsum: 30 kg
- Nemagon: 40 ml

Formula 10: Synthetic compost (Garcha 1980)

- Wheat straw: 300 kg
- Calcium Ammonium Nitrate: (CAN) 9 kg
- Urea: 3 kg
- Super phosphate: 3 kg
- Muriate of potash or potassium sulphate: 3 kg
- Wheat bran: 15 kg
- Molasses: 5 kg
- Gypsum: 30 kg
- Lindane dust: 250 g or linotex: 60 ml

Formula 11: Synthetic compost (Garcha and Kiran 1981)

- Wheat straw: 150 kg
- Paddy straw: 150 kg
- Calcium ammonium nitrate: 9 kg
- Urea: 3 kg
- Super phosphate: 3 kg

- Muriate of potash: 3 kg
- Wheat bran: 15 kg
- Molasses: 5 kg
- Gypsum: 30 kg
- Lindane dust or BHC (5%): 250 g

Formula 12: Synthetic compost (NCMRT, Solan, in Vijay and Gupta 1995)

- Wheat straw: 150 kg
- Paddy straw: 150 kg
- Calcium Ammonium Nitrate (CAN): 9 kg
- Wheat bran: 15 kg
- Urea: 4 kg
- Gypsum: 20 kg
- BHC (10%): 125 g

Formula 13: Synthetic compost (IIHR, Bangalore, in Vijay and Gupta 1995)

- Paddy straw: 150 kg
- Maize stalks: 150 kg
- Ammonium sulphate: 9 kg
- Super phosphate: 9 kg
- Urea: 4 kg
- Cotton seed meal: 5 kg
- Gypsum: 12 kg
- Calcium carbonate: 10 kg

Formula 14: Synthetic compost (IIHR, Bangalore in Vijay and Gupta 1995)

- Wheat straw 300: kg or paddy straw: 400 kg
- CAN: 9 kg
- Super phosphate: 9 kg
- Urea: 4 kg
- Wheat bran: 30 kg
- Gypsum: 12 kg
- Calcium carbonate: 10 kg

Formula 15: Synthetic compost for standard long method of composting (Mushroom Research Laboratory, Solan)

- Wheat straw: 300 kg
- Wheat bran: 30 kg
- Gypsum: 30 kg
- CAN: 9 kg
- Urea: 3.8 kg
- Muriate of potash: 2.5 kg
- Super phosphate: 3 kg

- Molasses: 5 kg
- Carbofuran (Furadan 3G): 183 g
- BHC (10%): 250 g or lindane dust (10%)

Formula 16: Synthetic compost for long method of composting (HRRS, Dhaulakuan)

- Wheat straw: 300 kg
- Wheat bran: 30 kg
- Gypsum: 30 kg
- CAN: 12 kg
- Urea: 4 kg
- Muriate of potash: 3 kg
- Super phosphate: 3 kg
- Molasses: 6 kg
- Cotton seed cake: 5 kg
- Mustard straw: 5 kg
- Carbofuran (Furadan 3G): 183 g
- BHC (10%): 250 g

Formula 17: Synthetic compost for long method of composting (in NRCM, Ahlawat and Verma 2000-01)

- Wheat straw: 1000 kg
- Wheat bran: 250 kg
- Urea: 18 kg
- Gypsum: 35 kg

(*ii*) *For short method of composting*

Formula 18: Straw compost: (Van lier et al. 1994)

- Chopped wheat straw: 250 kg
- Chicken (broiler) manure: 125 kg
- Gypsum: 20 kg
- Water: 1100 l

Formula 19: Synthetic compost (Seth 1978)

- Wheat straw: 1000 kg
- Chicken manure: 400 kg
- Brewer's grain: 72 kg
- Urea: 14.5 kg
- Gypsum: 30 kg

Formula 20: Synthetic compost (Garcha and Kiran 1981)

- Wheat straw: 300 kg
- Poultry litter: 60 kg
- Calcium ammonium nitrate (CAN): 6 kg
- Super phosphate: 3 kg
- Wheat bran: 15 kg

- Gypsum: 30 kg
- Lindane dust: 250 g or linotex: 60 ml

Formula 21: Synthetic compost (NCMRT, Solan, in Vijay and Gupta 1995)

- Wheat straw: 300 kg
- Wheat bran: 15 kg
- Chicken manure: 125 kg
- Urea: 5.5 kg
- Gypsum: 20 kg
- BHC (10%): 125 g

Formula 22: Synthetic compost (IIHR, Bangalore, in Vijay and Gupta 1995)

- Wheat straw: 300 kg
- Chicken manure: 120 kg
- Rice bran: 20.6 kg
- Brewer's grain: 22 kg
- Urea: 6 kg
- Cotton seed meal: 5 kg
- Gypsum: 10 kg

Formula 23: Synthetic compost (IIHR, Bangalore, in Vijay and Gupta 1995)

- Paddy straw: 3 tons
- Chicken manure: 1.5 tons
- Wheat bran: 125 kg
- Gypsum: 90 kg

Formula 24: Synthetic compost for standard short method of composting (Mushroom Research Laboratory, Solan, in Vijay and Gupta 1995)

- Wheat straw: 1000 kg
- Chicken manure: 400 kg
- Brewer's grain: 72 kg
- Urea: 14.5 kg
- Gypsum: 30 kg

Formula 25: Synthetic compost for organic mushroom extended (Phase-I) short method of composting (in NRCM, Dhar et al. 2007)

- Wheat straw: 1000 kg
- Poultry manure: 800 kg
- Brewer's grain (wet): 400 kg
- Wheat bran: 150 kg
- Cotton seed cake: 60 kg
- Gypsum: 35 kg
- Water: 3500–4000 l

Formula 26: Synthetic compost for short method of composting (in NRCM, Ahlawat and Verma 2000–2001)

- Wheat straw: 1000 kg
- Poultry manure: 500 kg
- Urea: 16 kg
- Wheat bran: 120 kg
- Gypsum: 30 kg

Formula 27: Synthetic compost (in Gupta et al. 2004)

- Wheat straw: 500 kg
- Poultry manure: 200 kg
- Rice bran: 50 kg
- Mustard oil cake: 25 kg
- Urea: 4 kg
- Gypsum: 40 kg
- Carbofuran (Furadan 3G): 183 g
- BHC (10%): 250 g

(iii) For indoor composting

Formula 28: (Vijay 2006)

- Wheat: 1 ton
- Poultry manure: 500 kg
- Wheat bran: 72 kg
- Urea: 15 kg
- Cotton seed cake: 20 kg
- Gypsum: 40 kg

Composting

There are mainly two methods for compost preparation. The first method is categorized as long method of composting which requires 28–30 days of time and does not require pasteurization, while, second is the short method of composting which requires 14–18 days of time and needs pasteurization.

(i) Long Method of Composting (LMC): The long method of composting is the oldest method in white button mushroom cultivation and this was first formulated with the use of empirical knowledge of growers. The use of that early method was not encouraging at beginning due to low yield and frequent crop failure. However, it became popularized as soon as the use of gypsum as an ingredient in the compost medium was started. Gypsum was effective to overcome the greasiness problem in the compost that was responsible for the poor growth of mushroom mycelia (Pizer 1937, Pizer and Thompson 1938). In India, the early method was modified by Mantel et al. (1972). They gave the first successful formula with the use of synthetic compost for growing white button mushroom (*A. bisporus*). Till date this method is very popular. In this method, there is no need of pasteurization for compost preparation, and hence, very little cost is involved in establishing a button mushroom production unit. However, still it has some drawbacks. It gives low yield and the production of mushrooms is uncertain due to improper composting which happens often at field level and causes crop failure. The long method of composting is completely done in an open place or in compost shed covered only at the top. The following steps of operation (with special reference to *formula No. 15*) are followed in this method.

Day (–2)—On this day, i.e. two days before stack preparation, the compost floor is washed with 2 per cent formalin. The straw is spread on the floor up to 9–12 cm height. Water is drenched into straw by running water through polythene pipe. Simultaneously, the straw is turned by frock for proper wetting. It is estimated that 5000 litres of water is required for one ton of dry straw. The watering should be done as much as it can absorb without run off. If any run-off water is there, it is to be collected from guddy pit and sprayed over the straw. The time to complete water soaking of wheat straw requires 48 hours before preparation of stack.

Day (–1)—On second day, that is the day before stack preparation, all other ingredients, except molasses, gypsum and pesticides are mixed separately and wetted with water by spraying. The wet ingredients are covered with moist gunny cloth and kept as such for next 24 hours.

Day (0)—The wet straw and other water soaked ingredients are mixed thoroughly with the forks. Now, they are piled on a floor grid fitted with side panels by attaching clumps and hooks. After filling completely the side panels are taken out to fix on another floor grid to prepare similar dice for preparing compost heap. Following this method, the size of the compost heap will be nearly 6′ in length, 5′ in breadth and 5′ in height. The breadth and height of compost heap are very crucial matter to maintain proper temperature and air inside the compost heap. In between two compost heaps, adequate space is kept for turning the compost. After completion of compost heap, it is kept as such for 5 days. During summer, if outer portion of straws are dried up sprinkling of water is done until it leaches from bottom. The leached water, if any, is collected from guddy pit and sprayed on the heap. The temperature of the compost pile increases up to 70 ± 2°C that should be checked after 2–3 days compost piling.

Day (6) (1st turning)—On this day the compost heap is broken and mixed with carbfuran (Furadon 3 G) @ 183 g per 300 kg wheat straw following the formula No. 15. Breaking of compost heap is done into three parts. The outer and top portions together are kept at one place, while, the middle and bottom portions separately at two different places. Now, these are after mixing with the nematicide as mentioned earlier, again restack in such a way that top and outer portions remain at bottom, the bottom portion remains at middle and the central portion remains at top and sides of the new stack.

Day (10) (2nd turning)—The compost stack is broken up into three parts as discussed earlier with 1st turning. The broken materials are mixed thoroughly with molasses (5 kg) diluting in a little water. These are turned again by preparing the compost stack with the top and side portions at bottom, the bottom portion at middle and the central portion at top and sides.

Day (13) (3rd turning)—The compost stack is break open and the composting substrates are mixed with gypsum (30 kg). These are now turned again and prepared the stack with the top and side portions at bottom, the bottom portion at middle and the central portion at top and sides.

Day (16) (4th turning)—The compost pile is break open and turned as followed in the preceding step. A little water may be required to add if the outer portions are dried up.
Day (19) (5th turning)—The compost pile is break open and turned as followed in the preceding step.
Day (22) (6th turning)—The compost pile is break open and turned as followed in the preceding step.
Day (25) (7th turning)—The compost pile is break open and the composting substrates are mixed with BHC powder (250 g) or lindane dust (10 per cent). These are again turned as followed in the preceding step.
Day (28) (8th turning)—The pile is break open and turned with a fork. These are kept open for next two days to remove excess ammonia formed during composting. After two days of keeping open, the compost is checked by smelling. If sweet smells appear the compost is ready for spawning. Otherwise, if the smell of ammonia still persists, then the compost is required to keep as such for one more day to release the ammonia. The good quality compost is dark brown in colour and smells sweet.

Note: *If chicken manure is used in any formulation of long method of composting, one more turning is required for releasing ammonia.*

(ii) Short Method of Composting (SMC): The short method of composting is the most important method in mushroom compost preparation. The quality of compost that is prepared by this method is of superior to that prepared by long method. Since in this method, the prepared compost is of more nutrient contents required for the growth of button mushrooms, which happened due to uniform aeration, better mixing of ingredients and accelerated microbial decomposition. Further, this compost is nearly free from harmful microbes. The formulation of short method was first described by Sinden and Hauser (1950, 1953). This method completes in two stages—the stage I (Phase I), that completes in outdoor condition and requires over a period of 7 days and stage II completes in an aerated room condition and requires over a period of 5 days including the period of steam pasteurization at 58–60°C. The steam pasteurization is an integral part in this method of composting (Lambert 1941). The short method of composting has revolutionized the mushroom industries all over the world, by producing more and assured crops with the use of unit amount of compost. In India, the short method with the use of wheat straw and chicken manure was formulated and standardized by Shandilya (1976) and Hayes and Shandilya (1977). Later the bulk pasteurization system was introduced by Tunney (1980). As per today's scenario, of Indian mushrooms, the method (SMC) as standardized by them is still popular in India. However, in using any formulation, the following points should be kept in mind. The total nitrogen content in the compost ingredients at initial stage will be 1.5–1.7 per cent on dry weight basis, which will increase to 2.1–2.3 per cent after completion of phase-I and to 2.2–2.3 per cent at the end of Phase-II. The moisture content during pile up in stack at beginning should be kept 75–77 per cent and at the end during filling in trays or bags be 68–70 per cent. The ammonia at the end of Phase-II should be below 0.1 ppm (Beyer 1999). To prepare compost in this method the following facilities are essential:

Composting yard—A composting yard of 100′ × 40′ is rsequired for a medium size mushroom farm.

Pasteurization facilities—The pasteurization may be done either in bulk pasteurization tunnel or in peak heating room.

Phase-I (Outdoor composting): At this stage all the required raw materials are mixed together and allowed to ferment under outdoor condition according to the following schedule:
Day (–4)—The floor of the compost yard is washed with 2 per cent formalin. The straw is spread on the floor up to 9–12 cm height. All the ingredients (e.g. as per the formula No. 24), except urea and gypsum are sprayed over the straw. These are watered and mixed together by frock. After mixing they are kept as such for 2 days.
Day (–2)—The wet substrate mixture is mixed thoroughly and turned by a fork. Simultaneously, water spraying is done until the substrate mixture becomes well saturated in wetting and that stopped just before the stage when further watering, if any, becomes excess and leached out. These are now gathered into small heaps of low height and kept as such for two more days.
Day (0)—The substrate mixtures which are kept in low height heaps are broken and mixed thoroughly with additional water and required amount of urea as per formula. These are now piled on a floor grid fitted with side panels and made compost heap of nearly 6′ in length, 5′ in breadth and 5′ in height and kept as such for next two days.
Day (2) (1st turning)—The compost pile is break open into three parts. The outer and top portions together are kept at one place, while the middle and bottom portions separately at two different places. These are now turned and pile up into the wooden frame to form the stack again with the top and side portions at bottom, the bottom portion at middle and the central portion at top and sides.
Day (4) (2nd turning)—The compost pile is break open and mixed with the required amount of gypsum. These are turned and restacked as followed in the preceding step.
Day (6) (3rd turning)—The compost pile is break open and mixed with a little amount of water. These are turned and restacked as followed in the preceding step.
Day (8th)—The compost is ready for indoor composting at phase II. At this stage conveyor belt may be used to transfer compost materials mechanically into bulk chamber (in large farm only), or manually may be transferred to bulk chamber or peak heating room. In last case, the compost is filled in trays.

Phase-II (Indoor composting): The phase-II process of compost preparation is completed in any of the following two different systems. It may be done **in peak heating room** where the compost materials are kept loosely in perforated plastic/wooden trays that are placed one after other keeping sufficient gaps in between for air circulation or in **bulk pasteurization chamber/tunnel** where the compost materials are kept in mass with 6–7 ft height on a grated floor. However, in both the cases compost is processed in two ways. In case of first way, which is known as pasteurization, the compost is pasteurized at high temperature (57–58°C) for killing the harmful microorganisms, while in second way, which is known as conditioning, the thermophilic microorganisms are allowed to grow at 45–50°C for converting ammonia into microbial protein. Later, during cultivation when the microbes become inactive at low temperature (25°C), this microbial protein is utilized by the mushroom. The compost temperature during pasteurization will never be above 60°C, since at that temperature beneficial thermophilic microbes will also die. Also it will never be 40°C during conditioning, since harmful mesophilic microbes may grow. The entire process of phase-II operation is completed within 7–8 days of time. During phase-II operation about 25–30 per cent compost weight is lost. For preparation of 20 tons of ready compost, 28 tons phase-I completed compost is required to fill in tunnel/ peak heating room and that may be achieved from 12 tons of raw materials.

In peak heating room—At first all the ventilation systems are closed in the peak heating room. The room is pre-warmed up to 40ºC by applying hot blower. Now, the trays are quickly filled with compost and kept inside the room in order to avoid temperature loss. The trays are placed one after other in such a way that there sufficient gaps in between two trays are present for air circulation. Now, the blower/circulating fan is kept open to circulate the air inside and this operation will be kept on all through the phase-II period. The temperature in the compost generally remains high during filling in trays. This temperature comes down due to air circulation. It is again brought down to 40ºC within 12 hrs and for doing this introduction of fresh air is required which can be achieved by operating ventilation device. Once the compost temperature reached to 40ºC, the ventilation system is closed. The compost temperature rises up in closed condition, but it is maintained at 40–50ºC for next two days by introducing fresh air with the aid of ventilation system, as and when required. After two days of time (24–36 hrs), live steam is introduced from boiler to increase the compost temperature and that is to be maintained 57–58ºC for about 6 hrs. Thereafter, the steam pipe is cut off and ventilation system is kept open to introduce fresh air. The air temperature inside the room is maintained to 40ºC, which is effective to maintain compost temperature at 45–55ºC for one day. After that, the compost temperature slightly comes down and that is maintained at 45–50ºC for about next 2–3 days or even more days, until the smell of ammonia goes out and sweet smell comes out from the compost on smelling. Thereafter, the temperature of compost is further cooled down to 25ºC for spawning in bag or in trays. The compost at this stage would be dark brown in colour with moisture content 66–68 per cent, pH 7.2–7.5, nitrogen content 2.0–2.3, fragile on pressing, without stickiness and no smell of ammonia persisted instead sweet smell coming out on smelling.

In bulk chamber/tunnel—In this system compost is placed loosely on the grated floor of bulk pasteurization tunnel. The filling of compost can be done manually or mechanically with the use of conveyor belt, in case mechanical filling uniform height is properly maintained. However, following any of the two methods of filling, the compost height is kept 6–7 ft. In the bulk chamber, there are several temperature sensors, of which one is kept below the plenum; while other two or three are kept inside the compost heap and two more in the air above the compost mass. As soon as the compost filling is over and the temperature sensors are fitted, the doors and the exit vents are made closed. The blower is started and that is kept running up during the entire period of phase-II. The blowing of fresh air and its recirculation via recirculation vent will bring uniform temperature in the compost and minimize the temperature difference (below 3ºC) between the compost and the inside air of the chamber. When the compost temperature reaches to 45ºC, the supply of fresh air is stopped. The temperature starts rising in the compost by self generating heat at the rate of 1.2ºC/h. At this stage, live steam is introduced from boiler to raise the compost temperature to 57–58ºC, which is achieved within 12 hours and, thereafter, that temperature is maintained for 6–8 hrs for pasteurization. After pasteurization, the temperature is brought down by slight introduction of fresh air and maintained at 45–50°C for about 3–4 days or till the period when the concentration of ammonia goes below 10 ppm (or differently below 0.1 ppm, Beyer 1999), while measured with dragger tubes (Vijay and Gupta 1995). Supply of steam is stopped, but it is required only when the temperature in tunnel goes below 45°C. This stage is called conditioning where only thermophilic microbes grow and convert ammonia into microbial protein. Now, after conditioning, the compost is allowed to cool down to 25°C by introduction of more fresh air. Now, the compost is ready for spawning, which has the characteristic of dark brown colour with moisture content 66–68 per cent, pH 7.2–7.5, nitrogen content 2.0–2.3, fragile on pressing, without stickiness and no smell of ammonia persisted instead sweet smell comes out.

(ii) Indoor composting: Apart from the above two methods recently one new method, i.e. indoor composting, is advocating by Directorate of Mushroom Research, Solan, Himachal Pradesh to follow. This has several advantages, it requires further less time (in all 12 days only) than short method, it emits very less amounts of harmful gases and facilitates favourable condition more to grow thermophilic fungi, like *Scytalidium thermophilum*, *Humicola insolens* and *Humicola grisea*, etc. This method requires less space, and reduction and loss of raw materials is also less (Vijay 2006).

In preparation of the compost all the ingredients as in above formula are mixed thoroughly in dry form. These ingredients are properly wetted so that it contains 70 per cent moisture. Run off water is regularly collected and sprinkled over the wetted straw. These wet substrate materials are spread over the composting yard (around 8–10″ height) and are trampled hard by running a front loading machine (Bob cat) several times to increase the bulk density of the materials and to shred the wheat straw. Water is sprinkled over the substrate materials while bob cat machine is running over them. This time the moisture is maintained to 74–75 per cent. Now, these substrate materials are piled up to make a heap and this is kept for 48 hrs as undisturbed.

After 48 hrs of incubation, the stack is broken up and the materials are placed in phase-I tunnel (bunker, with two sides above 7–8 ft height are open) to make a heap of 5–6 ft height. Three temperature sensors are placed one each at bottom, middle and top of the compost stack inside the tunnel. Now, this is kept as it is for overnight for fermentation and raising temperature at centre to 50–55°C. During that time temperature at bottom and top remains 45–50°C. Now, the blower fan starts for 5 minutes per hour and this is regulated by fixing timer this condition is maintained for 36 hours and during this period the temperature of compost at centre goes up to 62–68°C and at top and bottom it remains 48–52°C. After this stage, the compost stack is broken and turned to pile up in stack in the same tunnel. Similar procedure in blower operation is followed for next 36 hrs as done in the preceding stage. Total period for completion of phase-I in this method requires 6 days. Next stage is phase-II in bulk chamber and that procedure is same as in short method of composting (Vijay 2006). After 6 days of phase-II stage the ammonia in compost is checked by smelling or analysis and that should remain below 10 ppm (Vijay and Gupta 1995). In case it remains high, one or two more days may be required for conditioning.

CULTIVATION OF WHITE BUTTON MUSHROOM (*Agaricus bisporus*)

Cultivation of white button mushrooms (*A. bisporus*) is a complex process and it requires technical skill for growing successful crop. The mushroom is heterotrophic for carbon compounds and during cultivation the action of other living organisms (microbes) is required to provide suitable complex organic compounds for its growth, i.e. as compost. Its mycelial growth (vegetative stage) occurs at 22–28°C with optimum on 24–25°C, while for fruit body development (reproductive stage) it requires 12–18°C (optimum 16–18°C). The relative humidity of 85–95 per cent and proper ventilation during fruit body formation are also required for cropping.

Cultivation on Shelf

The shelf cultivation is very common in India. This is done by the seasonal growers of Haryana and Punjab who cultivate button mushrooms on shelf prepared by spreading polythene sheet on bamboo structure. In mechanized mushroom farms, also shelf cultivation is used. In this system, four to five tiers of shelves made of wooden framework or angle iron are used for holding shelf board of about 2.4 m long and 90 cm broad. The shelves are prepared as such that the lowermost shelf remains 20 cm above ground and the space between two shelves is 60 cm. The compost with 66–68 per cent moisture is suitable for shelf cultivation. The spawn mixed compost (with spawn 0.5–0.75 per cent) is spread over the shelves up to height of 15–20 cm and for this about 80–100 kg of compost is required per square metre of mushroom bed. It takes 15 days for mycelial growth and thereafter, the myceliated compost is covered by pasteurized casing soil up to 3–4 cm height. During spawn running stage ventilation is kept minimum for creating the condition of higher CO_2 (0.1–0.5 per cent) contents in air favouring mycelial growth (San Antonio and Thomas 1972). The temperature inside the room is maintained 23 ± 2°C and the relative humidity is 85–90 per cent. This condition is kept up to 7 days after casing for case run. The casing layer at that time is kept covered with moist old newspaper to avoid direct watering. After case run the moist paper is taken out and the ventilation system is maximized to introduce fresh air inside the room, which increases oxygen level and decreases CO_2 level. The temperature is brought down to 15–17°C, while RH is maintained at pin head formation stage at 90–95 per cent and during fruit body development at 80–85 per cent. This condition triggers the transformation of vegetative stage to reproductive stage. The pinheads of mushroom are appeared within 7–10 days of venting, i.e. airing. The mushroom takes 16–22 days for first harvest after the application of casing material and that is harvested for a period of next 6 weeks.

Cultivation in Bags

The ready compost with around 65 per cent of moisture is mixed with 0.5–0.75 per cent spawn of mushroom and filled in polythene bags @ 10 kg/bag. The compost depth is kept 30–35 cm in the bag. Now, the bags are kept on rack for a fortnight in a mushroom growing house. Alternatively, in shake up spawning method, the spawn is introduced later into 5 cm depth of compost with the help of thumb pressure after filling in bag. After 7 days (differently

8–9 days) of incubation, the mouth fold of the bag is made open and a vigorous shake is given to the bag and mixed the half grown mycelia with compost. This is again closed and kept for another 7 days. During cultivation the temperature of the room is maintained 23 ± 2°C, while relative humidity (RH) is 90–95 per cent and CO_2 as accumulated in the room. Air handling instrument is kept on for circulation of air inside the room in order to maintain uniform temperature. After 15 days of spawn run, the mouth of the bag is opened. Now, the myceliated compost is covered with 3–4 cm thick steam pasteurized casing layer. This casing layer is covered with a wet paper to avoid direct watering on layer. A spray of light formalin solution (0.5 per cent) is spread over the bed to avoid contamination (Vijay and Gupta 1995). This covering paper is kept moist for 7 days for case run, i.e. ramification of mycelia in casing material. On completion of case run, the wet paper is taken out. Soon, fresh air is introduced inside the room to bring the CO_2 level lower. The temperature in room is maintained to 15–17°C, while RH to 85 per cent. The above condition triggers the transformation of vegetative stage to reproductive stage. The pinheads of mushroom appear within 7–10 days of venting, i.e. airing. The mushroom takes 16–22 days for first harvest after the application of casing material and that produces in flushes for a period of next 6 weeks (Figure 6.4).

Figure 6.4 Button mushroom cultivation in bags. **A.** White strain, **B.** Brown strain.

Casing and Casing Material

Casing material is a thin layer of covering on actively growing mushroom at its vegetative phase, i.e. mycelia. This is applied to check the growth of mycelia and triggers to shift into reproductive phase. The casing material should be nutritionally poor with high water holding capacity, pH 7.0 to 7.8 and low electrical conductivity. Further, it should favour the growth of actinomycetes and bacteria which have positive correlations with the yield of button mushroom in bed (Jarial and Shandilya 2004).

Preparation of casing soil/material

Casing material: The best material for casing is peat moss which grows in certain temperate regions. In India, peat moss of good quality is not available in sufficient quantity. So, other materials are used for casing and for that different formulations of ingredients have been appeared to make the casing material in button mushroom cultivation. Some of these are as follows:

1. Spent mushroom substrate (compost, 2 years old) + FYM (1–2 years old) + Clay loam (2:1:1)
2. FYM (1–2 years old) + Coir pith (coir industry waste) (1:1 v/v ratio)
3. Spent mushroom substrate (compost, 2 years old) + Coir pith (coir industry waste) (1:1 v/v ratio)
4. Spent mushroom substrate + sand (2:1)
5. FYM (1–2 years old) + loam soil (1:1) (mostly used)
6. Garden loam soil + sand mixture (4:1 v/v ratio)
7. Spent mushroom substrate (compost, 2 years old) + Sand + Lime (4:1:1 v/v/v)
8. Spent compost + Dung compost (1:2 ratio) (Tirkey et al. 2009)

Processing: In order to avoid salt accumulation in casing materials during use, the materials are washed and soaked in running water in a tank for about one day. Thereafter, excess water is drained out. Now, these materials are disinfected either chemically or with steam.

Chemical disinfection

Chemical disinfection is done by fumigating with 2 per cent formaldehyde (2 litres of formalin (formaldehyde 40 per cent) diluted in 40 litres of water). In practice, the casing soil is spread over a concrete floor to a height of 15 cm and drenched by the formalin solution @ 3 litres per cubic metre of casing soil. The casing soil is then piled up to make a heap and covered that with a plastic sheet for 48–96 hrs. Later, the cover is opened and the heap of casing material is broken into a thin layer for removal of chemicals in air. This operation is done at least two weeks before the use to provide sufficient time for removing excess moisture and formalin, since traces of later one hinder mycelial growth of mushrooms.

Steam pasteurization

This is the most advanced method for disinfecting the casing material/soil. In this method the casing materials are filled in wooden/plastic trays and placed one after other on the grated floor in casing pasteurization chamber. Sufficient gaps between trays are kept for air circulation. Electronic temperature probes thermometers are placed in casing soil for check the temperature from outside the chamber. The door is closed. The blower is kept on and live steam is introduced for recirculation of air and steam in order to increase temperature in casing soil. When the temperature is reached to 70ºC then it regulates by air handling unit (inserting/cutting off fresh air) to maintain temperature 65–70ºC for about 5–6 hrs. Thereafter, it is cooled to 25–30ºC and used in bed for casing.

Crop Management

Watering

Sprinkling of small amount of water on mushroom bed and on the walls and floor of mushroom house is done regularly during cropping period to maintain required bed moisture, wetness of paper that covering casing layer and room humidity. During first flash, which generally occurs after 22–24 days after casing, the moisture of compost and casing soil is kept high for developing more mushrooms per unit area. However, watering on the casing soil is done by sprinkler method only before initiation of pinhead. During pinning no spraying is practiced, since, watering during pinhead formation causes bacterial blotch and discolouration of fruit body. When the pinhead formation is over and the mushroom button size grows to pea seed shaped, then water is spread for its development. Water spraying is minimized after second flash onwards in the successive flashes which come after 6–8 days of intervals.

Air circulation and venting

Forced air circulation is essential during cropping period for distribution of heat and CO_2 which generated during cropping and to evaporate water humidity all through the room. Venting is required to introduce fresh air for minimizing room temperature and CO_2 concentration to a required limit.

Prevention of disease and pest attacks

Quite a large number of diseases of all fungal (dry bubble, wet bubble, cobweb, etc.), bacterial (bacterial blotch) and viral origins affect white button mushrooms, in addition to the attacks of different kind of weed fungi which spoil compost quality. This subject has been discussed elaborately in Chapter 8. In the present chapter the preventive measures as practiced during cultivation are highlighted. The diseases, especially fungal and bacterial ones, and the weed fungi are more prevalent when the compost prepared in long method is used for cultivation. So, to avoid their occurrence, the compost of long method is treated with carbendazim (0.015 per cent) and formalin (0.5 per cent) 1–2 days prior spawning, thus, requiring 1.5 litres of formalin and 50 g of carbendazim for the compost prepared from 3 quintal of base material. Alternatively, the compost is disinfected by solarization for about 6–8 hrs. For this purpose, the compost is spread in sun light as thin film and covered with polythene sheets for checking evaporation of moisture and heat transmission.

Light

Light is not required for the growth of this mushroom. However, it is essential to carry out different operations.

Harvesting and Packing

Harvesting of white button mushroom is done when the mushroom is at button stage with diameter of cap is 2.0–2.5 cm, the pileus is still rolled and closed, the veil is intact and the lamellae are not visible. The mushroom is hold by fingers and gently pulls upwards with simultaneous twisting at base. A small knife is used to cut the basal attachment of soil and mycelia. Soon after plucking, a small amount of casing soil is placed in the gap which formed after harvesting and pressed that for leveling. The fruit bodies are packed in perforated paper packets or perforated polythene packets with 200 or 400 g of fresh mushrooms. For long transportation, insulated ice containers are used in which mushrooms can be transported for a period of 8–10 hrs. Mushrooms are sold fresh as soon as possible after harvesting, since, they loose original colour and perish, with time. The mushrooms can be kept in refrigerator (4°C) for two days. For long time preservation canning can also be done after blanching. During harvesting if diseased mushrooms are noticed, they are taken out separately and at the diseased spot common salt (NaCl) is applied to cover the diseased spot. Mushrooms can also be harvested at cup stage when their diameter is 2.5 cm with ruptured veil and open lamellae, however, their demand is lesser than the button ones. After harvesting every flush, bed is cleaned by removing the damaged parts and late appeared pinheads of mushrooms and sprayed with 0.5 per cent formalin.

CULTIVATION OF TEMPERATURE TOLERANT WHITE BUTTON MUSHROOM (*Agaricus bitorquis*)

The temperature tolerant white button mushroom is with silky white cap of rather coarse and short stipe. It requires 4–5°C higher temperature than common white button mushroom for its growing. Thus, bacterial blotch and other fungal diseases are very common if the mushroom is not cultivated in proper hygienic conditions. However, the mushroom is resistant to viral diseases (Dieleman-van Zaayen 1972).

Compost

The mushroom required good quality of compost, any compost prepared in long method of composting is not suitable for growing this mushroom. So, always short-term composting is followed.

Formulation of compost

Any compost ingredients required for short method of composting and suitable for the growth of temperate button mushrooms may be used for cultivation of subtropical button mushroom. However, two formulas which specifically suggested for growing the temperature tolerant white button mushrooms are given below.

Formula 1: Synthetic compost (NRCM, Solan)

- Wheat straw: 1000 kg
- Poultry manure: 500 kg
- Urea: 14.6 kg
- Wheat bran: 50 kg/brewers grain: 72 kg
- Gypsum: 30 kg

Formula 2: Synthetic compost (NRCM, Solan)

- Paddy straw: 1000 kg
- Poultry manure: 500 kg
- Urea: 14.6 kg
- Wheat bran: 50 kg/brewers grain: 72 kg
- Gypsum: 30 kg

Composting

Composting in short method, which completed in two phases—phase-I (outdoor) and phase-II (indoor), is always practiced for cultivation of temperature tolerant white button mushroom cultivation. The quality of compost prepared by this method is superior to that of long method. Since in this method, compost is prepared under uniform aerated condition, with provision of better mixing of ingredients and accelerating microbial decomposition. Further, this compost is nearly free from harmful microbes due to pasteurization at high temperature.

Cultivation

Cultivation on shelf

The shelf cultivation is done by the large mushroom growers in mechanized mushroom farm. In this system, four to five tiers of shelves which are made up of wooden frame work or angle iron are used to hold up the shelf board. The shelves (about 1.5 m wide) are prepared as such that the lowermost shelf remains 20 cm above ground and the space between two shelves is maintained 60 cm. The compost with 66–68 per cent moisture is suitable for shelf cultivation. The spawn mixed compost (spawn 0.5–0.75 per cent) is spread over the shelves up to a height of 15–20 cm and covered with a thin polythene sheet/paper. About 80–100 kg of compost is used per square metre of mushroom bed. The spawned substrate is kept for 15 days to grow mycelia and, thereafter, the plastic cover is opened and the myceliated compost is covered by pasteurized casing soil up to a height of 3–4 cm. During spawn running stage, insertion of fresh air is not needed, since the mushrooms can tolerate high amount of CO_2 and that condition favours the mycelial growth (Hasselbach and Musters 1971). The temperature inside the room is maintained to 28–30°C and the relative humidity to 85–90 per cent. This condition is kept all through the spawn run stage, i.e. up to 14–16 days after spawning. The mycelia of *A. bitorquis* is much finer, therefore, after completion of spawn run the compost looks less white in comparison to *A. bisporus.* However, a few weeks after casing, the compost turns into complete white (Vedder 1978). After completion of spawn run, the polythene cover is made open and a casing of 4–5 cm thickness is applied with the pasteurized casing material. The casing soil is made by mixing two years old Farm Yard Manure (FYM) and loam soil in 4:1 ratio. Alternatively, it can be prepared with two-year old decomposed spent compost alone. Prior the use of both FYM and spent compost is leached in water for 4–6 hrs to remove the accumulated salts. Soon after covering with casing material, the casing layer is spread thoroughly by a solution of 2 per cent formalin and 0.5 per cent benlate mixture to disinfect and prevent mold attacks. The casing layer is now kept as covered with moist old newspaper to avoid direct watering. After the case run for a period of 7–8 days, the moist paper is taken out and the ventilation system is made on to introduce fresh air inside the room which increases oxygen level and decreases CO_2 level. The temperature in the room is brought down to 24–26°C, while RH is maintained 90–95 per cent. This condition triggers the transformation of pin heads. The pin heads of mushroom appear within 7–10 days of venting, i.e. airing. The mushroom takes 16–22 days for first harvest after the application of casing material and that produces in flushes for a period of next 6 weeks. During fruit body development, i.e. soon after the fruit body attaining pea seed size, the RH inside the room is maintained to 80–85 per cent.

Cultivation in bag

The ready compost (with about 65 per cent of moisture) is mixed with 0.5–0.75 per cent spawn of mushroom and filled in polythene bag @ 10 kg/bag. The compost depth is kept 30–37 cm in the bag. Now, the bags are closed by folding the upper part and kept on a rack in mushroom growing house for a fortnight. Alternatively, in shake up spawning method, the spawn may be introduced later after filling the compost in bag into a depth of 5 cm with the help of thumb pressure. After 7 days (differently 8–9 days) of incubation, the mouth fold of the bag is made open and a vigorous shake is given to the bag, and thus, the half grown mycelia is mixed with the non-colonized compost. This is again closed and kept for another 7 days. During cultivation, the temperature of the room is maintained to 28–30ºC, while the relative humidity is 90–95 per cent and CO_2 as accumulated in the room. Air handling instrument is kept on for circulation of air inside the room in maintaining uniform temperature. After 15–16 days of spawn run, the mouth of the bag is made open. The myceliated compost is now cased with 4–5 cm thick steam pasteurized casing layer. The casing material is prepared with the mixture of two years old Farm Yard Manure (FYM) and loam soil in the ratio of 4:1 or with two year old decomposed spent compost alone. Now, the casing layer is covered with a wet paper to avoid direct watering on the layer. A spray of light formalin solution (0.1 per cent) mixed with benomyl/carbendazim (0.05 per cent) is spread over the bed to avoid contamination (Vedder 1978, Vijay and Gupta 1995). This casing layer is kept moist for 7–8 days for case run, i.e. ramification of mycelia in casing material. Now, the wet paper is taken out and fresh air is forcefully introduced inside the room with blower. This fresh air brings down the CO_2 level. The room temperature is maintained to 24–26ºC, while RH to 90–95 per cent. This condition triggers the transformation of vegetative stage to reproductive stage. The pin heads of mushroom appear within 10–12 days of venting. In this mushroom cultivation mass pinning is a problem, so the entry of fresh air is restricted intermittently for moderate pinning. After appearance of pin heads, the RH in room is maintained to 85 per cent. The mushroom takes 22–26 days for first harvest after the application of casing material and thereafter, it is harvested for a period of next 6–8 weeks in flushes at 10–12 days interval.

Casing and Casing Material

Casing is done with 4–5 cm thick layer of casing material on the actively growing mushroom at its vegetative phase, i.e. mycelia. This is applied to check the growth of mycelia and triggers to shift the vegetative mycelia into reproductive phase. The casing material should be nutritionally poor, with high water holding capacity, pH 6.8 to 7.2 and low electrical conductivity.

Preparation of casing soil/material

Casing material

1. Spent mushroom substrate (compost, 2 years old)
2. FYM (1–2 years old) + loam soil (4:1) (mostly used)

Processing: In order to leach the accumulated salt in casing materials, the materials are washed with and soaked in running water in a tank for 4–6 hrs. Thereafter, the excess water is drained out and the materials are disinfected either chemically or by steam.

Chemical disinfection

Chemical disinfection is done by fumigating with 2 per cent formaldehyde (2 litres of formalin (formaldehyde 40 per cent) diluted in 40 litres of water). The casing soil is spread over the concrete floor up to a height of 15 cm and drenched by the chemical solution @ 3 litres per cubic metre of casing soil. The casing soil is piled-up to make a heap and covered with a plastic sheet for 48–96 hrs. Later, the cover is opened and the heap of casing material is broken into a thin layer for removal of chemicals in air. This operation is done at least two weeks before the use to provide sufficient time for removing excess moisture and formalin, since the traces of latter one hinder mycelial growth of mushrooms.

Steam pasteurization

This is the most advanced method for disinfecting the casing material. In this method the casing materials are filled in wooden/plastic trays and placed one after other on the grated floor in casing pasteurization chamber. Sufficient gaps between trays are kept for air circulation. Sensors of electronic thermometers are placed inside the casing material for checking the temperature from outside the chamber. The door is made close. The blower is kept on and live steam is introduced for recirculation of air and steam in order to increase temperature in casing soil. When the temperature reached to 70°C then it regulates by air handling unit (inserting/cutting off fresh air) to maintain temperature 65–70°C for about 8 hrs. Thereafter, it is cooled to 25–30°C and used in bed for casing.

Crop Management

Watering

No watering is given on mushroom bed before application of casing soil. However, sprinkling of water on the walls and floor of mushroom house is done regularly to maintain required room humidity. Watering on bed is done by sprinkler method on the paper that covered the casing layer before initiation of pinhead. During pinning no spraying is practiced, since, watering during pinhead formation causes bacterial blotch and discolouration of fruit body. When the pin head formation is over and the mushroom button size reaches to pea seed shape, then water is applied for its growth. Frequency of water spraying is minimized after second flush onwards in the successive flushes which come after 10–12 days of intervals.

Air circulation and venting

Forced air circulation is essential during the cropping period for distribution of heat and CO_2 that generated during cropping and for evaporation of water to raise humidity to about 85 per cent all through the room. Venting by air handling operation is required to enter fresh air for minimizing temperature and CO_2 to the required limits, i.e. 24–26°C and below 1500 ppm, respectively.

Disease and pest management

After application of casing layer light solution of formaldehyde (0.1 per cent) is spread over the casing soil to avoid contamination. In certain cases benomyl/carbendazim (0.05 per cent) is also sprayed along with formaldehyde solution for preventing diseases and weed fungi. During casing, the entry of flies/insects from outside is prevented by fitting fly mesh (28–30 mesh net) on doors and vents. The casing soil is also drenched with malathion if the flies at larval stage attacked mushrooms during cropping period. Fumigation of DDVP is also done to kill flies.

Light

Light is not required for the mushroom growth. However, to carry out different operations light is essential.

Harvesting and Packing

Harvesting of this mushroom is done when it reached button stage with diameter of cap is 3.0–4.0 cm, the pileus is still rolled and closed, the veil is intact and the lamellae are not visible. The fruit body is hold by fingers and gently pulled upwards with simultaneous twisting at base. A small knife is used to cut the basal attachment of soil and mycelia. Simultaneously, a small amount of casing soil is placed in the gap as created after harvesting the mushroom and pressed for levelling. The mushrooms are packed in perforated paper packet or perforated polythene packet with 200 or 400 g of fresh mushrooms. The mushrooms are chilled for 2–4 hrs at 5–6°C in order to avoid anaerobic rotting and marketed thereafter as fresh. For long distance transportation, this mushroom is superior to common button mushroom, since at 15°C it can be kept fresh for a few days. Insulated ice containers are used to transport mushroom for a period of 2–3 days. The mushroom in refrigerator (4°C) can be kept for two to three days. For long time preservation canning can also be done after blanching. During harvesting, if diseased mushroom is noticed then that is taken out separately and on the diseased spot common salt (NaCl) is applied to cover that. Mushroom can also be harvested at cup stage with ruptured veil and open lamellae. However, its demand is less than the button ones. After harvesting of every flash, the bed is cleaned by removing the damaged parts and late appeared pin heads of mushrooms and sprayed with 0.5 per cent formalin.

Improvement in Cultivation Method

Three different methods of improvement are suggested for cultivation of both kinds of button mushrooms. These are done at the time of casing as follows:

Supplementation of soybean seed meal: Improved yield with the supplementation of soybean seed meal was first observed by the Dutch farmers and later that method has been adopted by many growers all over the world. In this system after completion of spawn run, the coarsely powdered soybean meal which disinfected with 0.5 per cent formalin are mixed with the spawn run compost @ 1 kg/quintal of compost and covered with the casing material. It increases yield by 15–24 per cent (Dhar 1995).

CAC'ing (spawned casing)

In this system, the casing materials are mixed with a small amount of spawn (grain spawn or compost spawn) and applied on the bed as casing layer. In doing so, grain spawn is used @ 1 per cent of casing material, while compost spawn @ 1.5 kg/m^2 bed area (Gupta et al. 1989, Gupta and Dhar 1993).

Ruffling of spawn run compost

In this method, the completely colonized/myceliated compost is broken before casing to disturb the mycelia. The broken myceliated compost is levelled and cased immediately. This breaking favours faster growth of mycelia (Gupta and Dhar 1992).

Note: *The term 'CAC'ing' is used while mixing of colonized compost is being done with casing material.*

CHAPTER

7 Cultivation of Specialty Mushrooms

OYSTER MUSHROOM CULTIVATION

Oyster mushroom is most widely cultivated in China and South Korea. It is also flourishing in almost all the mushrooms growing tropical countries, including India, rapidly. In India, it is mainly cultivated in the eastern and southern provinces. The cultivation of this mushroom is very easy and can be grown by any farmer on almost all agricultural residues. However, most commonly rice straw or wheat straw is used for its cultivation. The paddy straw of *aman* rice is more suitable for its cultivation (Biswas and Singh 2004). The straw does not require any procession to use as substrate. Simply water soaking and disinfection either by chemicals, steam or hot water are enough to make the substrate ready for its cultivation. As per the global production, it ranked third with the production of 876,000 metric ton/year, sharing 14.2 per cent of total mushroom production in 1997 (Chang 1999). Thereafter, its production has increased in many folds. In 2003 China alone has produced 2,488,000 metric ton, which ranked first in the country and shared 24.0 per cent of total mushroom production in that country (Chang 2007). Not only in China, in other East Asian countries also, this mushroom is widely cultivated. In South Korea its production is highest with 61,965 metric ton, production/year (based on 2003 data). In India, the mushroom is cultivated mainly in small farms or in farmer's houses as cottage industry. However, nowadays, it has gained much popularity due to its easy method of cultivation and ability to grow on various waste materials with high production efficiency. This mushroom offers good prospects of its commercialization due to longer shelf-life, pleasant flavour and excellent nutritional value. Accordingly, in certain places, its commercial cultivation has already been started (Prakash and Tejeswini 1991, Bakakrishnan and Nair 1995). Much progress has also been made on the improvement of production technology with the use of various agricultural residues, cultivation technologies and strains in different parts of our country (Bano and Srivastava 1962, Jandaik and Kapoor 1974, Vijay and Sohi 1987, Bahukhandi 1990, Chandra et al. 1995, Biswas and Singh 2004, 2007, 2008, 2011).

Mushroom Species

The name oyster has been given to the fungi of *Pleurotus* sp. on the basis of their shapes like oyster shells. In certain places, these mushrooms are called as *dhingri* mushroom. These are white wood rotting fungi. In nature the mushrooms grow on lignocellulosic materials, like dead parts of plants, wood logs or tree stumps. The mushrooms are used as food for their delicious flavour and other culinary properties. Several species of *Pleurotus*, such as, *Pleurotus sajor-caju, P. flabellatus, P. florida, P. ostreatus, P. citrinopileatus, P. cornucopiae, P. sapidus, P. membranaceous, P. eryngii, P. fosssulatus, P. eous, P. djamor* and *P. platypus* are brought under artificial cultivation (Figure 7.1). Among the species, *Pleurotus sajor-caju, P. flabellatus, P. sapidus, P. membranaceous, P. citrinopileatus* and *P. eous* prefer higher temperature for their growth and can be grown in almost all areas in India during different times (Table 7.1). The rest of the species required comparatively low temperature and can be grown during winter.

Figure 7.1 Different species of oyster mushrooms. **A.** *Pleurotus sajor-caju*, **B** and **D.** *P. florida*, **C.** *P. flabellatus*

Table 7.1 Temperature Requirement of Different Oyster Mushroom Species for their Growth

Oyster Mushroom Species	*Optimum Temperature Required for Growth*	
	Mycelial Growth (Vegetative)	*Fruit Body Formation (Reproductive)*
Pleurotus citrinopileatus	25–30	22–28
Pleurotus cornucopiae	25–30 (25–30)	18–22 (20–30)
Pleurotus eous	25–30	22–26
Pleurotus eryngii	18–22 (25)	14–18 (13–18)

(*Contd.*)

Table 7.1 Temperature Requirement of Different Oyster Mushroom Species for their Growth (*Contd.*)

Oyster Mushroom Species	*Optimum Temperature Required for Growth*	
	Mycelial Growth (Vegetative)	*Fruit Body Formation (Reproductive)*
Pleurotus flabellatus	25–30	22–26
Pleurotus florida	25–30 (25)	18–22 (15–25)
Pleurotus fosssulatus	18–22	16–20
Pleurotus membranaceous	25–30	22–29
Pleurotus ostreatus	25–30 (25)	20–22 (10–17)
Pleurotus sajor-caju	25–30 (25)	22–26 (18–25)
Pleurotus sapidus	25–30	22–26

Morphology

Pleurotus cornucopiae

The mushroom grows in large cluster on dead tree trunks or on dead portion of living trunk. It has a floury-spermatic odour. Its pileus is 4–10 cm (differently 5.1–12.7 cm) broad, convex initially, later depressed and margin folded downwards. Its colour may be dun-ochreous or pallid-whitish. It is firm, smooth, margin often splitting. The stipe is central or eccentric, not lateral, striate where the decurrent and often anastomosing lamellae extend down the stem. Gills are subdistant, decurrent, usually 2.5–5.1 cm long, white, smooth, firm, solid, in tufts and several joined at base. Flesh is white. Basidiospore is oblong, 8–11 μm × 3.5–5.0 μm. Spore in mass is of pale lilac colour (Zadrazil 1978, Purkayastha and Chandra 1985).

Pleurotus eous

The fruit bodies usually grow on dead portion of living trees and on paddy straw. Sporophore is caespitose, imbricate, usually sessile. Pileus is up to 9.0 cm in diameter, initially spathulate, pink or flesh coloured, flabelliform when old, glabrous, margin incurved. Gills are crowded, decurrent, whitish or creamish, narrow, thin, lamellulate of four different lengths. Flesh is 0.25 cm thick, white, fleshy to brittle, hyphae interwoven, thin walled, with clamps. Odour is mild. Basidia are 4 spored, narrowly clavate. Basidiospores are hyaline, cylindrical, thin walled, size 6.0–8.0 × 2.5–3.5 μm. Spore print is white, Pleurocystidia are absent (Purkayastha and Chandra 1985).

Pleurotus eryngii

This mushroom is known as 'king oyster'. It requires cod shock for fruit body formation. Its mycelial growth is very slow, It is susceptible to diseases. Sporophore usually grows on dead roots, centrally stipitate. One fruit body while cultivated may be of 300–400 g. Pileus is 4.0–15.0 cm broad, at first convex, later expanded and depressed, gray brown, reddish brown, dirty yellow or rusty tawny, margin folded downwards. Gills are subdistant, decurrent, broad, whitish to cinerous, soft. Stipe is whitish, 3–10 cm (differently 10–14 cm) long, base fusiform.

Flesh is white, tough. Odour is mild. Basidiospore is white, ovate to oblong, granulose, 8–11 μm × 4–5 μm (differently 7.0–9.0 μm × 3.5 μm). Spore in mass is white to liliac gray (Zadrazil 1978, Purkayastha and Chandra 1985, Kong 2004).

Pleurotus flabellatus

Sporophores are short stipitate growing on dead tree trunks, ground or decaying logs. Base of sporophore is sponge like. Pileus is slender, flabelliform (fan shaped), white or turning red or pink, at first tomentose, later becoming smooth, attenuated downward. Gills are decurrent narrow. Stipe is short, tomentose. Hymenophoral trama are irregular. Basidiospores are cylindrical, hyaline, punctuate, non-amyloid. Pleurocystidia are absent. Cheilocystidia are present (Purkayastha and Chandra 1985).

Pleurotus florida

This mushroom is widespread in temperate, subtropical and tropical regions. It is now confirmed that *P. florida* and *P. ostreatus* represent a single species (Gonzalez and Labarère 2000). However, this mushroom is some what smaller and finer in structure and different in colour, which changes with temperature (Zadrazil 1978). There are two groups of *P. florida* at the subspecies level. One group is sexually compatible with *P. ostreatus* and the other with *P. pulmonarius* (Kong 2004). Sporophores are usually growing in clusters on dead tree trunks or branches and rarely on living trees, usually hygrophanous, at low temperature the colour is light brown, but pale at higher temperature (Kong 2004), large. Pileus is 8–20 cm broad, spathulate to kidney shaped, white, grey, light brown (at low temperature), pale (at high temperature) or sometimes yellowish after drying, surface smooth, margin incurved. Gills are not crowded, decurrent, anastamosing at the base, white, yellowish when dry, broad. Stipe is eccentric or lateral, 1.0–3.0 cm long, 0.5–2.0 cm thick, farm, sometimes hairy at the base. Flesh is white, soft, spongy, 0.5–1.5 cm thick near the stipe. Taste and smell are pleasant. Hymenophoral trama are irregular. Basidia are 4 spored, 30.0–38.0 × 6.0 μm. Basidiospores are white, oblong, 7.0–10.0 μm long. Spore print is lilac.

Pleurotus fossulatus

Sporophores usually grow on dead wood or roots of plants, stipitate, fleshy, soft, tough and rigid when dry. Pileus is 3.0–13.0 cm broad, sometimes up to 28.0 cm, plano-convex when young, whitish to creamish, surface initially glabrous, then stribiliform squamulose and rimose when mature, margin involute when young, but revolute when old. Gills are crowded, decurrent, anastomosing towards the stipe, whitish, up to 1.3 cm broad. Stipe is eccentric lateral, rarely central, up to 7.0 cm long and 3.0–6.0 cm broad, white, attenuated downwards, solid, surface rugulose. Flesh is white, up to 3.5 cm thick, brittle on drying, hyphae loosely interwoven, thin walled, with clamp. Basidia are 4 spored, clavate. Basidiospores are hyaline, ellipsoid or cylindrical, some times globose, thin walled, guttulate, 9.0–13.0 × 4.5–6.0 μm. Cheilocystidia are many, hyaline, lanceolate to clavate (Purkayastha and Chandra 1985).

Pleurotus ostreatus

Sporophores are usually growing in clusters on dead tree trunks or branches and rarely on living trees, widespread in the temperate zones, usually hygrophanous, grey to grayish brown, large. Pileus is stemmed at the side, 8–20 cm broad, shell shaped, spathulate, tongue shaped or kidney shaped, gray, gray brown or slate-gray in colour, surface smooth, margin incurved. Gills are not crowded, whitish or grey, decurrent, anastamosing at the base, white, yellowish when dry, broad. Stipe is short, eccentric or lateral, 1.0–3.0 cm long, 0.5–2.0 cm thick, farm, sometimes hairy at the base. Flesh white, soft, spongy, 0.5–1.5 cm thick near the stipe. Taste and smell are pleasant. Hymenophoral trama are irregular. Basidia are 4 spored, 30.0–38.0 × 6.0 μm. Basidiospores are white, oblong, 8.0–12.0 μm long and 3–4 μm broad. Spore print is lilac (Zadrazil 1978, Purkayastha and Chandra 1985).

Pleurotus platypus

Sporophore is caespitose, stipitate, eccentric to subcentral. Pileus is 1.5–4.0 cm broad, convex initially, later depressed, sometimes sub-infundibuliform, tawny brown, smooth, pileal surface is with a non-gelatinized cutis of repent, gills are moderately crowded, white. Stipe is solid, light coloured, 1.5–5.0 cm long, base bulbous-clavate, wrinkled. Flesh is up to 0.3 cm thick. Basidia are clavate, basidiospores are hyaline, cylindrical, thin walled, 6.5–9.0 × 2.5–3.5 μm. Pleurocystidia is absent (Purkayastha and Chandra 1985).

Pleurotus sajor-caju

Sporophores are solitary or in groups, occurring on decaying plant parts. Pileus is 5–14 cm in diameter, oyster shell shaped to deeply infundibuliform, often lobed and folded at maturity giving a coralloid appearance, white to gray or dull brown coloured intense towards the margin, surface smooth, margin irregular and curved. Gills are distinctly formed, decurrent, white when fresh and yellow when dry, thin edge smooth. Stipe is eccentric, usually 1–3 cm long, rigid on drying. Basidia are clavate 35–55 × 5–12 μm. Hymenophoral trama are irregular. Basidiospores are hyaline, cylindrical, smooth, non-amyloid, point of attachment with sterigmata prominent, 6.5–7.5 × 2.7–3.3 μm. Spore print is white. Generative hyphae are thin walled with clamp connections. Skeletal hyphae are thick walled. Ligative hyphae are thick walled with wide lumen (Purkayastha and Chandra 1985).

Facilities and Materials Required

1. Mushroom house or shed (partitioned into two halves for spawn running and mushroom production separately)
2. Substrate store house or shed
3. Cemented tank with outlet/plastic drum with outlet (for soaking straw in water or in disinfectant solution)
4. Water supply
5. Wooden, bamboo or steel racks
6. Hand chopper/chaff cutter-1 (required for cutting hard substrate materials and straws)
7. Metallic drum for water boiling (required if disinfection of substrate is to be done by hot water or vapour treatment)
8. Wire cage (required if disinfection of substrate is to be done by vapour treatment)
9. Plastic bucket: 2
10. Bamboo cane basket: 2
11. Wooden plank

12. Sprayer
13. Gunny bags
14. Fire wood or large gas oven (required if disinfection of substrate is to be done by hot water or vapour treatment)
15. Polypropylene bags (16″ × 20″)
16. Polythene sheet (transparent)
17. Nylon thread
18. Nylon ropes of about 0.75 cm diameter (required if oyster mushroom is cultivated on hanging bags)
19. Mosquito net
20. Substrates, any one of the agricultural residues, like fresh paddy straw of *aman* paddy, wheat straw, pea haulms, pea pod shell, maize stalks, etc. or sawdust and rice bran.
21. Spawn, planting spawn
22. Fungicide: Carbendazim (Bavistin), mancozeb (Uthane M-45) and zineb (Uthane Z-78)
23. Antibiotics: Streptomycin sulphate and tetracycline hydrochloride mixture (plantamycin, pushamycin or agrimycin)
24. Insecticides: Malathion (if fly attacks occurred), endosulfan (optional, may be used as traces during water soaking for preventing rat and fly attacks)
25. Disinfectant: Bleaching powder, formalin

Mushroom House

Any type of house as used for dwelling purpose is suitable for oyster mushroom cultivation. Thus, one can grow these mushrooms in small scale at a corner of a house. Only special care which required is that during sporulation stage the mushroom beds are to be covered by a plastic or cloth sheet to avoid spore allergy. However, for commercial cultivation, separate thatched house with mud or bamboo mat wall plastered by cement and sand mixture, and cemented floor is well suited. The house is divided into two halves for spawn running and mushroom growing separately. Further, for commercial cultivation a separate store shed is required for storing of substrates well in advance.

Rack: Wooden or steel rack (breadth 3 ft, height 8 ft, shelves 4, gap between two shelves 2 ft, length 18 ft or as convenient to keep gap 2 ft between the rack and wall and that of 3 ft between two racks) with lowermost shelf 1 ft above the ground is used for cultivation of mushroom. Alternatively, bamboo or wooden supports at top are used to hang mushroom bags in chain with the support of four medium thick nylon ropes which knotted together just above each bag.

Tank: Cemented tank with outlet at bottom is needed. This is convenient to prepare on the back of mushroom house for soaking the substrates in disinfectant solution or in plain water for over night (16–18 hrs).

Substrate Preparation

Substrate types

Wheat straw and paddy straw are the conventional substrates to grow this mushroom. Particularly, paddy straw of *aman* season is the best substrate for cultivating oyster mushrooms. However, the other agricultural residues, such as, mustard stick, *toria* stick, sesame residues with pod shell, arhar pod shell, pea haulms, pea pod shell, french bean pod shell, black gram/green gram pod shell, hulled maize cob and maize stalk (if fresh) as well as sawdust of saw mill supplementing with rice bran are more or less equally good substrates for oyster mushroom cultivation. Pea haulms give more yields within shorter period in certain species. In case of sawdust use as substrate, supplementation with commercial rice bran (*kuro*) @ 20 per cent of dry weight is required to obtain standard yield (Biswas and Singh 2007, 2010). The combinations of different agricultural residues are also used as substrates. For example, the combinations of soybean straw and linseed straw, and linseed straw and *Ageratum* twigs are found suitable in oyster mushroom (*P. sajor-caju*) cultivation (Kumar et al. 2004).

Processing of substrates

Substrates like paddy straw, maize stalks, mustard sticks, sesame residues and haulms of agricultural crops are cut into 5–8 cm long pieces with a hand chopper or chaff cutter. However, to save labour, the crumpled paddy straw and soft haulms of agricultural crops may be used directly without cutting. Wheat straw is readily available as pieces after thrashing, thus, it does not require any further procession. In case of maize stocks, sesame residues with pod shell and mustard/toria sticks cutting is essential due to their hardness. Further, for maize stalks cutting in fresh is convenient since they become very hard and difficult to chop after drying. Sawdust collecting from saw mill is used directly without any further procession. However, in case of fresh sawdust it is kept dry in store house for about half a year before using as mushroom substrate.

Water soaking and disinfection

Water soaking and disinfection of substrates are done by various ways as convenient according to the available facilities. Here three methods which commonly used and can be done by any farmers at their homes are described.

1. **Hot water treatment:** In this method substrate is water soaked for 16–18 hrs for overnight in a cemented tank or any large container. On next day morning, the substrate is taken out and kept on a cemented platform or in a bamboo cane basket in heap for about 1 hr to remove the excess water from the substrate. After one hour of water removing, the water soaked substrate is kept in a large vessel/bucket and there boiled water is poured as such that the substrate is fully immerged in hot water. The whole set is covered by a wooden plank and kept for 2 hrs. Time to time boiled water may be needed to add for maintaining the temperature of about 65°C or more. The substrate is then taken out again in a bamboo cane basket for draining out excess water and later spread in sun to remove excess moisture for about 1–2 hrs depending upon the brightness of sun light. Thus, the moisture is kept 60–70 per cent in substrate, which can be visibly identified when the surface layer of substrate appears as dry. In other way, the chopped straw is soaked in water as usual and boiled in a container for about 30 minutes. Then, the excess water is drained out and cooled the straws by spreading on a clean surface before using for mushroom bed preparation.

2. **Steam pasteurization:** Steam pasteurization is normally done in a peak heating room or in bulk chamber both facilitated with boiler and blower as discussed in the preceding chapter.

However, alternatively for simplification, a top open metallic drum is placed on a brick made oven and filled in water up to 1/5th portion by pouring two buckets of water. The water is then boiled by burning fire wood or gas at bottom. The overnight soaked substrate after removing water is put in a wire net cage which has four stands and placed inside the drum as such that the water level remains below the bottom of the cage. The drum is then covered by a lid with a small opening at centre or by water soaked two gunny bags placing cross-wise. The steam coming out from the boiling water passes through the substrates and finally through the opening of lid or leakages of gunny bags. After, 45–60 minutes of steaming, the cage along with substrate is taken out and the substrate is sprayed out in sun for removing excess water.

3. **Chemical treatment:** Chemical disinfection is the easiest method and suggested to follow for minimizing time, labour and expenditure on appliances and fire woods. Moreover, mushroom production efficacy of chemically treated substrate is more than that done by the other treatments. In this method, a chemical solution is prepared according to the composition of carbendazim 3 g (which comes 75 ppm in the solution), formalin 25 ml (500 ppm) and water 20 litre (Vijay and Sohi 1987). The substrate is immersed in that solution for about 16–18 hrs. On completion, the solution is drained out and the substrate is spread in sun for removing excess moisture. Alternatively, the chemical disinfection is done by water soaking the substrate for 16–18 hrs, draining out the excess water and dipping that in the chemical solution for half an hour, successively. Later, the excess solution is drained out in basket and excess moisture is removed in sun. In case of sawdust using as substrate, plastic buckets are more convenient for chemical treatment. A mosquito net is placed over the bamboo cane basket to hold the smaller particles during decanting of the excess solution.

Note: *The substrate treated with steam pasteurization or autoclaving at* 30 lbs p.s.i. *(not discussed) is reported to give more production (Bahukhandi 1990). Paddy straw substrate treated with lime solution* (1.0%) *for overnight* (16–18 hrs) *is a non-hazardous chemical measure equally effective for mushroom production and minimizing contaminants (Biswas,. S. 2011, unpublished).*

Cultivation

On paddy straw/wheat straw and other soft agricultural residues

Bag preparation in poly bag method: Mushroom beds are prepared in various ways, but the most convenient way is that in which bed is prepared in a polythene bag (Figure 7.2). In this method, transparent polythene bag (20″ × 16″) is perforated (hole size 0.4″ diameter) at a distance of 4″ all over its body surface. Disinfected paddy straw (about 1.5 kg on dry weight basis) is made ready after any of the earlier methods of disinfection and a bottle/bag of spawn (150 g) is used to prepare a bed. The substrate is placed inside the bag in 4–5 layers by pressing with hand palm. After completion of one layer 30–40 g of spawn grains are spread over the substrate layer. Then, another layer of substrate is prepared and the spawn grains sprayed over that similarly. In this way 4/5th portion of the transparent polythene bag is filled up by the spawned substrate. Now, the bag is closed tightly by tying with a piece of nylon thread and incubated in spawn running room in dark for about 18 days. In case of small particle-substrate, as for saw dust, the spawn and substrate are mixed thoroughly and filled in the bag. Later, the mouth of the bag is closed by tying with a piece of nylon thread and incubated in dark at room temperature (25–28°C), depending upon the temperature requirement of the cultivated species.

Figure 7.2 Cultivation of oyster mushrooms in bags. **A.** Disinfection, **B.** Drying, **C.** Poly bag filling, **D.** Spawn running, **E.** Fruit body production, **F.** Harvesting and weighing.

Bed preparation in cube bed method: This is also very common method in case of oyster mushroom cultivation. In this method a perforated plastic sheet is placed over a plain surface. A wooden frame measuring 50 cm l × 25 cm b × 8 cm h is placed over the polythene sheet. Cut straws (alternatively intact straws) after disinfection and removal of excess moisture are put inside the wooden frame and pressed on by a wooden lid of somewhat smaller size than the inner side of the frame. A lump full spawn grains are spread over the substrate after taking out the lid. In this way 4–5 layers are prepared and spawned on each layer and even that on top layer. After completion, the frame is taking out by pressing the substrate with the lid and pulling the frame upwards. The polythene sheet is wrapped over the substrate well. A few pieces of nylon threads are used to tie it firmly. Now, the bed is incubated at dark in a room for spawn running.

Spawn running: The spawned substrate in closed polythene bags or cube beds wrapped by polythene sheet is kept in a cool and dark place for 18–20 days for spawn running. Periodical observation is done to see the spawn growth.

Fruit body production: After completion of spawn running, which is visible from outside the transparent polythene sheet as white mycelial growth with brownish tinge, the polythene bag is cut open (bag method) or polythene sheet is unwrapped from bed (cube method). The bag or bed is kept on racks or on ropes as hanging in chains with a support above in the mushroom house, where humidity has been artificially raised by hanging wet gunny cloths at different places. Sprinkling of water regularly twice at morning and afternoon is done to develop fruit bodies. Fruit bodies develop within 3–4 days of bag opening. Thereafter, the mushrooms grow in the same bed for two to three times in flushes at about 7–8 days interval. At last, when the productivity is decreased enough, the mushroom bed is thrown out in a compost pit for manure preparation.

Note: *In certain cases, polythene bags are used without any perforation. Here, the mouth of the bags are loosely tied, closed by folding or kept open. In some other cases complete opening of polythene bag is not practiced after completion of spawn run. Here, only perforation is made at different places after completion of spawn running in substrate. However, in the latter case the yield is found lower than the fully opened bag.*

Harvesting: Harvesting of mushrooms is generally done at morning when the fruit body matures which is identified by seeing the edges of the caps that started to fold or curl upwards. This is done by giving a gentle twist at the base of fruit body. The adhering substrate particles are removed by hand picking or cutting and the mushrooms are made ready for marketing or food preparation.

On sawdust

Fresh sawdust is not suitable for oyster mushroom cultivation. Sawdust collected from saw mill is to be dried and kept for 6–8 months in room condition for killing the living cell and drying the substrate materials. After the said period, the saw dust becomes ready for use in mushroom cultivation. Gunny bag filled sawdust is immerged in disinfecting solution as described earlier or in water for about 18 hrs. The bag is taken out from water and kept on a raised platform for an hour. If, that is immerged in plain water, then the sawdust is placed in a big plastic bucket for disinfection in hot water, which at boiling stage are poured in the sawdust filled bucket and covered by a wooden plank. Alternatively, steam pasteurization can also be done as described earlier in this chapter. Later, the excess water is decanted and sawdust is kept in sun for an hour for drying up to the stage when its surface is looked like drying. Rice bran (*kuro*) or wheat bran (*bhushi*) weighed about 15–20 per cent of the dry sawdust substrate is also soaked for about 18 hrs in a little amount of disinfecting solution or water on the day before use, i.e. this is done on the same day when sawdust soaking installed. The excess moisture of rice bran or wheat bran is also dried in sun for an hour. If only water alone is used for steeping the bran, then that is to be sterilized at 15 lbs p.s.i. for 15 minutes in an autoclave. Now, both the substrate ingredients are mixed thoroughly. The spawn @ 4–5 per cent of the dry substrate is also mixed thoroughly with the substrate mixture. This is now filled in perforated polythene packet and kept in dark for spawn running. When the white mycelia covered the substrate, upper half of the bag is cut open by a sharp blade. This is now kept in production room for fruit body formation as usual as done for straw substrate bag (Figure 7.3). Further, it is to be remembered that saw dusts from all woody plants are not good substrates. For example, sawdust of *Trema orientalis* is very good substrate, while that of *Gmelina arborea* is very poor (Biswas and Singh 2008).

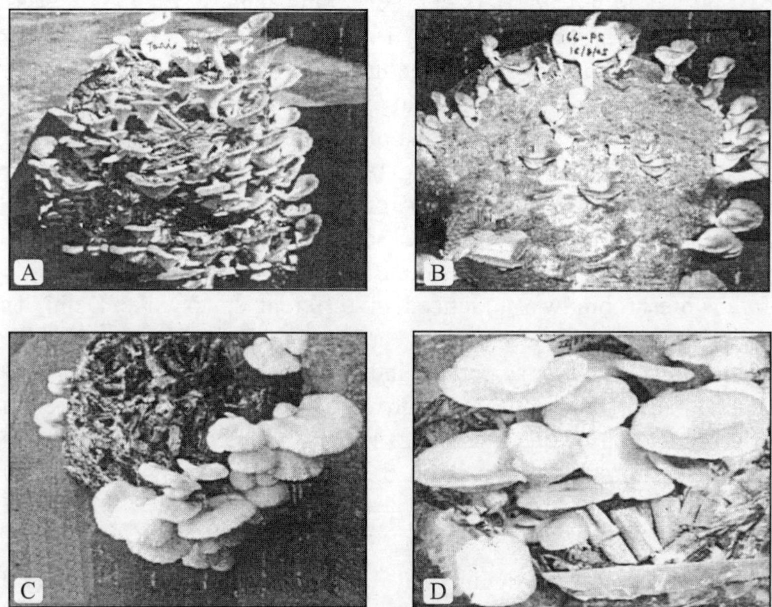

Figure 7.3 Oyster mushrooms on hard agricultural residues and sawdust. **A.** Toria stricks, **B.** Sawdust, **C.** Brinjal twigs, **D.** Maize stalk.

On hard agricultural residues

The hard agricultural residues, like sesame residue, maize stalk and brinjal twigs, are needed to cut into small pieces. Maize stalk and brinjal twigs are cut when fresh for easiness to piece them. In case of mustard and toria sticks and sesame residues dry residues are only available after harvest of crops from the pods. All these materials are to be kept dry for about 4–6 months before use. Rest of the cultivation procedures, like disinfection of substrate, removal of moisture, spawning, spawn running, production of fruit bodies and harvesting are same as described with the cultivation on paddy straw substrate.

Improvement in Cultivation Methods

Certain improvements in yield on straw based substrate can be made by supplementation with the wheat bran (Bahukhandi 1990). Improvement in yield of oyster mushrooms can also be made by hybridization of different strains of a particular species (Bahukhandi and Munjal 1990).

PADDY STRAW MUSHROOM CULTIVATION

The paddy straw mushroom belongs to the genus *Volvariella* under family Pluteaceae of Basidiomycotina. It is also known as straw mushroom and Chinese mushroom. It has hundreds of species, subspecies and varieties, however, only 13 species have been reported from India (Rath 1962, Ghosh et al. 1967). Of the 13 species only three namely *Volvariella volvacea* (Bull. Ex. Fr.) Sing., *V. diplasia* (Berk and Br.) Sing. and *V. esculenta* (Mass.) Sing. are reported to be cultivated in India (Roy et al. 1978). However, the separate entity of the species is doubtful due to absence of noticeable antigenic differences in the precipitin tests and presence of similarity in morphological characteristics among the cultivated strains. Thus, it is concluded that all cultivated forms of paddy straw mushroom in the world are the members of *V. volvacea* (Singer 1975, Chang 1978). The paddy straw mushroom cultivation was started in China about 200 years ago. Around 1932–35 this mushroom cultivation was introduced in Philippines, Malaysia and other south-east Asian countries. Its cultivation was first introduced in India at Coimbatore (Tamil Nadu) by Thomas et al. (1943). Later, the cultivation of this mushroom was practiced in different states, like Delhi, Uttar Pradesh, Orissa, Punjab, West Bengal, etc. However, at present, its cultivation is still restricted in small farms or in farmer's houses as cottage industry in Orissa and West Bengal (Suman and Sharma 2005). The paddy straw mushroom grows at a relatively high temperature of around 35°C. The world production of this mushroom is 181,000 metric ton/year sharing only 3 per cent of total mushroom production to about 6,160,800 metric tons of 1997 in the world.

Morphological and Other Characters

Sporophores are centrally stipitate, umbonate and medium but occasionally large in size (Figure 7.4). Pileus is 8–24 cm wide, smooth, brown, gray or dark brown in colour, occasionally whitish under in-house cultivation, dark brown or dark grey at umbo, but light gray or light brown near margin, at first campanulate then expanded to umbonate, fleshy, soft fatty or silky to touch, grayish radial coloured streaks present near the middle of the pileus, margin regular, but sometimes splits, cuticle easily peelable, veil absent. Gills are crowded, distinctly formed, free, thin, white at elongation stage, brown at mature stage, broad, narrowed towards margin, straight, margin complete. Stipe is central, cylindrical, attenuated upwards, 8–14 cm long, whitish, ending below with a solid bulbous base covered with dark gray coloured volva, bilobed and straight at elongation stage, on maturation becomes crumpled and soft, persistent. Flesh is white, soft. Hymenophoral trama are inversed. Basidia are clavate, tetrasterigmatic, 18.7–29.9 × 6.8–12.5 μm. Basidiospores are oval to ovoid, sometimes obovoid with hilar appendage, smooth, thin walled, 6.8–9.3 × 4.5–6.5 μm. Pleurocystidia are lanceolate to clavate, 47.6–62.9 × 13.6–18.7 μm. Cheilocystidia are usually ventricose rostrate. Spore print is salmon pink coloured on white paper. No clamp connection is in mycelia (Chang 1978, Purkayastha and Chandra 1985, Biswas 1988).

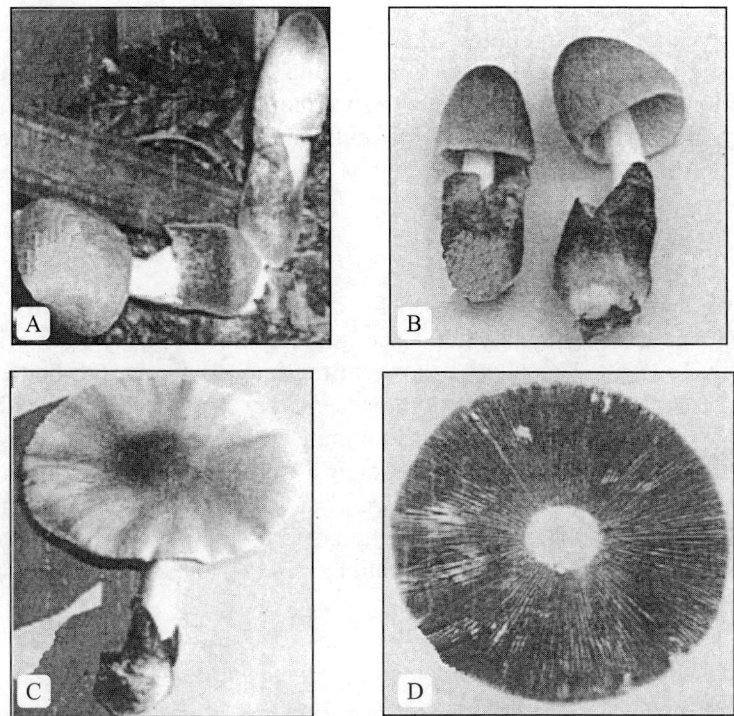

Figure 7.4 Vovariella volvacea with gray to gray-brown sporocarps. **A.** On pea pod shell substate, **B.** Just opening state after harvest, **C.** Fully matured fruit body and **D.** Spore print on white paper.

Facilities and Materials Required

The following facilities and materials are required for cultivation of paddy straw mushroom.

Mushroom house

The mushroom house must be well ventilated and with enough light inside the room. The mushroom generally grows better near the open window. The verandah of any house or top covered shed is suitable for cultivation of this mushroom in conventional methods. Mushroom house should be attached with water tank for soaking paddy straw for 18 hours and a slanted platform for removing excess water from the water soaked straw. In large scale production closed shade with fluorescent illuminated condition is used. In such case the mushroom house is made up on bamboo/wooden frame and covered inside with polythene plastic sheet of 0.4 mm film, while outside by polyform sheet of 1.27 cm for insulation. The house is fitted with a ½ hp electric blower, a polyethylene air duct for ventilation and four windows, two on each long side of the house.

Substrate store house or shed

Alongside the mushroom house, a shed is needed for storing the substrates, like, paddy straw. In case of compost preparation a composting yard may be prepared with a cemented floor covered with tin.

Substrate

One year old paddy straw (aman) is suitable for cultivation of this mushroom. In industrial areas where cotton waste is available, the fresh cotton waste is suitable for preparation of compost. Paddy straw mushroom can also be cultivated on various agricultural wastes. The important ones are palm bunch waste, pea halms, pea pod shell, rajmash pod shell (Promod et al. 2004, Biswas and Singh 2008).

Spawn

Both straw spawn and grain spawn are used for cultivation of mushrooms. The straw spawn is prepared by mixing the overnight water soaked small pieces of paddy straw with 10 per cent gram dal powder. The mixed medium is now stuffed into a wide mouth bottle (milk bottle) for sterilization, inoculation and incubation, successively. For grain spawn wheat or paddy grain is used. The grains are boiled for 25–30 minutes then they are allowed to remove extra water on a strainer or netted container. The grains are now mixed with calcium carbonate (1–2 per cent) and calcium sulphate (3–6 per cent) and stuffed into a bottle or polypropylene packet for sterilization, inoculation and incubation, successively. After 3 weeks of incubation, the spawn is used for mushroom bed preparation.

Gram dal powder

Gram dal powder is required as supplements during spawning @ 300 g/16 kg straw in case of conventional system.

Rack

Wooden or bamboo rack is required to support the mushroom bed prepared by straw. The rack should be 1 m of breadth in case of conventional cultivation. In each bed four raised poles are made at four corners for covering with polythene sheet.

Polythene sheet

This is required for covering the mushroom bed keeping 6″ gap between bed and sheet in order to raise humidity surrounding the bed.

Water supply connection

Water is needed to soak substrate, to clean mushroom house and watering in bed.

Sprayer

A hand sprayer is required for spraying fungicides, insecticides and antibiotics to manage diseases and pests.

Chemicals

Chemicals like formaldehyde (disinfectant), carbendazim (fungicide), mancozeb/zineb (fungicide), agrimycin/plantamycin (antibiotic) and endosulfan (insecticide) are required.

Lime

The lime is required if compost is prepared.

Rice bran

Rice bran is required if compost is prepared.

Chicken manure

Chicken manure is required if cotton waste compost preparation is done.

Wooden frame

Wooden frame of 90 cm × 90 cm × 30 cm is required for compost preparation.

Mini boiler

Mini boiler is required if pasteurization is done of the partially decomposed compost.

Cultivation

The paddy straw mushroom is mainly cultivated on two types of substrates, like, paddy straw and partially decomposed cotton waste. The yield performance on cotton waste is more than that of paddy straw. Cotton waste is mainly used in large scale production of this mushroom at the industrialized areas in Hong Kong, China and other South East Asian countries. In India, cotton waste is not easily available every where, thus, most of the cultivation is practiced on rice straw based substrate. The paddy straw is used for both outdoor and indoor cultivation of this mushroom. Several, agricultural wastes, like, sugarcane bagasse, dried banana leaves, wheat straw, oat straw, wood waste, pineapple waste, jute waste, etc. were also used for cultivating this mushroom, but none of them substantiates the yield of paddy straw. Several techniques are used for the cultivation of this mushroom in the tropical areas which thrive in the temperature range of 30–35°C and R.H. of 75–85 per cent (Quimo et al. 1990, Biswas and Singh 2008a).

Traditional crisscross bed method

In India straw mushroom is cultivated during summer and rainy months. The months starting from April to October are suitable for cultivation of this mushroom. The favourable period for growing this mushroom is June to October (Biswas and Singh 2009).

In traditional crisscross bed method, the intact paddy straw is used without any disinfection. Clean, fresh, dried and uncrumpled paddy straw is required for good crops. The straw is generally marketed as bundles; two such bundles are tied together at bottom end and leafy end. The unequal parts are cut off. Now, this bundle weighs about 0.5–1 kg and its length is 3 ft. These bundles are tied together to form a larger bundle and immerged in a fresh water tank for 18 hrs. On next day they are taken out from the tank and kept on a slanted platform for about 2–3 hrs for decanting excess water and removing excess moisture. Now, these bundles (with about 70 per cent moisture) are used for mushroom bed preparation. For preparation of one mushroom bed 32 straw bundles, two bottles/packets of spawn and 300 gram dal powder/arhar dal powder are needed (Munjal 1975, Purkayastha et al. 1980, 1981a, 1981b). At first, eight bundles are used for preparation of first layer. The bundles are laid on a raised wooden/bamboo platform in such a way that the bottom end and top end of the bundles remain alternatively in successive manner. The layer is now spawned around the periphery in a line by inserting a small amount of spawn pushing inside the bundle by finger pressure and keeping six inch gap towards the edge. Small amount of gram dal powder is also spread over the spawn. The second layer is prepared as like as first layer but here the straw bundles are placed in opposite direction, i.e. in crisscross manner. The spawning and spreading of gram dal powder on 2nd layer is done as like as done on first layer. The third layer is placed in opposite direction to the second layer, but in the same direction to the first layer. Spawning on this layer is done by broadcasting the spawn all over the surface. Similarly, gram dal powder is spread all over the spawn surfaces. The fourth layer is placed in opposite direction to the third layer, but in the same direction to the second layer. This layer acts as covering layer and spawning is not done on the 4th layer. After placing the fourth layer, the whole bed is tied at two places by ropes with simultaneous pressing by hands on the bed. The mushroom bed is now covered with a thin transparent polythene sheet keeping 6 inches gap between the polythene sheet and the mushroom bed. Regular watering is done after 4–5 days of spawning onwards. Pinheads of mushrooms appear from all the sides and top of the bed on 6–7 days after spawning. The mushrooms are harvested at button, egg and elongation stages on 10th or 11th days.

Modified crisscross bed method

Bundle preparation and disinfection: In case of modified method of crisscross bed preparation, 1 year old crumpled paddy straws are tied into bundles of about 15 cm diameter and 75 cm length (Biswas and Singh 2008b). Twenty such bundles are used for preparation of a bed. The paddy straw bundles are water soaked in a tank over night (for 18–20 hrs), thereafter; these are taken out and kept for an hour on a slanted platform for draining out the excess water. After removal of excess water, the straw bundles are dipped into the chemical solution, prepared according to the composition of carbendazim 3 g, formalin 25 ml and water 20 litres, for about 10 minutes for disinfection. The disinfected straw bundles are kept in a heap on the slanted platform for about 2–3 hrs to remove the excess water again.

Bed preparation and cultivation: Five straw bundles are used in one layer on a raised structure at about 1.5' height, prepared by wood or bamboo strips keeping gaps of 2 inches between strips. Spawning is done by pushing the spawn grains inside the straws around the periphery of the first layer in a line on 4–6 cm inner from the edge. Now, dusting of gram dal powder on the spawn is done. The next layer of straw bundles is prepared by keeping the straws in just opposite direction of those of the first layer in crisscross manner. Spawning and dusting of gram dal powder are done as earlier. The straws of the 3rd layer are placed in opposite direction of the second layer. On the third layer, spawning and dusting of gram dal powder are made throughout the surface. Fourth layer is prepared by placing the straws in the same direction of second layer. This layer is the covering layer and spawning is not required. Now, the whole bed is pressed and fastened tightly with coir/nylon threads. For preparation of such bed, six–eight kg of paddy straws (on dry weight basis), two packets (300 g) bottles of spawn and 200 g of gram dal powder are required. The complete bed is now covered with a transparent polythene sheet, keeping 10–15 cm gap above and around the bed and the covering reaches up to 10–15 cm below the lowermost layer. Water is sprayed as and when the outer straws showed dryness. Mushroom fruit bodies come out within 10–12 days after the spawning. Two more flushes of mushrooms may be obtained at 7–8 days interval.

Note: *Instead of gram dal powder, rice bran (5 per cent of straw weight) or sterile jaggery (2 per cent of the dry paddy straw) can be sprayed over spawn after spawning (Kaur* et al. *2004, Sangeetha* et al. *2008).*

Cube bed method

Bed preparation and cultivation: Four nylon threads are arranged in crisscross manner on a plain and clean surface, keeping some gaps between threads. Thereafter, a wooden frame, measuring 1 ft × 1 ft × 1 ft is placed on the ropes. Wet and disinfected crumpled straw, as made ready by earlier methods except bundle preparation, are put inside the frame and pressed to form a layer with a lid of smaller size than the inside space of the frame (Figure 7.5). Now, bits of spawn grains are sprayed over the straw layer. A little amount of gram dal powder is also sprayed over the spawned layer. This is done repeatedly to form 4–5 layers. The topmost layer is the covering layer and kept as unspawned. Now, the lid is placed on the top layer and pressed on the substrate along with pulling upwards the frame. After taking out the frame and the lid, the bed is tied with the 4 nylon threads on pressing with hand palms. For preparation of this type of bed, 3 kg paddy straw (cut to make larger pieces or intact), one packet (150 g) of spawn and 100 g of gram dal powder are required. Five bamboo strip pieces of 1 ft length and with one end sharp are inserted half of their length inside the bed at four corners as inclined and at the middle of top as straight. The bed is kept on a bamboo or wooden rack where enough day-light is present and covered the bed by a transparent polythene sheet, keeping six inches gap between the sheet and bed with the support of five inserted bamboo strips. The lower portion of the bed is kept uncovered. The watering to the bed is not generally required. However, if required by seeing the dryness of straws, may be sprinkled over the straw surface. On 7th day onwards pin heads of mushroom fruit bodies appeared and on 10–12th day they become ready for harvest. (Biswas and Singh 2009, 2010).

Figure 7.5 Cultivation of paddy straw mushroom.

Outdoor cultivation

Several technologies are used for the production of paddy straw mushroom under outdoor condition. However, the most common practices are as follows:

Bed system

The beds as prepared in crisscross manner can be placed on a raised bamboo made platform under a shady place. In order to protect from heavy rain, the beds are covered by straw mulches or polythene sheets but these are removed immediately after rain. Watering is started a few days (3–4 days) after spawning. Fruit bodies develop from all sides on 10–12 days.

Spreading paddy straw system: This cultivation is practiced at the places where paddy straw mushroom grows naturally during favourable season. In this method, paddy straws of 1–2 years old are sprayed in a layer on the ground in home stead. During the optimum favourable condition the fruit bodies are come out from the straw which attached with the soil. During first harvest in the season, a few of the mushroom sporophores are crushed in water and that mixture is spread over the straw for further flush. By this way two to three flushes are taken out and the decomposed straw is used for manure preparation.

Improved cultivation method/modern cultivation technology

This method is practiced in industrial areas of Hong Kong, China, etc. where cotton wastes are abundant. Cotton waste alone or cotton waste and rice straw mixture (1:1 ratio) is used for compost preparation. Compost preparation is being done with the use of different formulas.

Compost preparation: It can be done by three methods, discussed as follows:

(i) Method 1—In China, cotton waste is used. Different grades of cotton wastes are mixed and used for preparation of compost. The cotton wastes are moistened with sprinkling water and gradually mixed with limestone (2 per cent) and chicken manure (5 per cent). Now, the mixture is put inside a wooden frame (90 cm × 90 cm × 30 cm). When the frame is completely filled with the moist cotton wastes mixture, it is pulled up and similarly another layer of moist cotton waste mixture is placed on the first layer. By this way, the height of the stack of cotton waste mixture is maintained to 70–90 cm. The stack is covered with plastic sheet and kept as it is for 2 days. After two days, the compost is turned upside down and inside out with simultaneous supplementation of rice bran @ 2 per cent and kept for another two days in stack preparing with the use of wooden frame and covering with plastic sheet for fermentation. After, completion of latter two days, the compost is filled in netted platforms/racks and maintained compost height about 10 cm, breadth 90 cm and length of any size above 90 cm for pasteurization and mushroom bed preparation.

(ii) Method 2—Cotton waste and rice straw (1:1) are separately sprinkled with water for wetting one day (day-0). Next day the wetted substrates are piled up by mixing together with 1–4 per cent lime and stack up to a height of 1.5 m. The stack is covered with a polythene sheet and left for 2 days for fermentation. After two days, i.e. on day-3, the compost is turned and mixed with 4–5 per cent rice bran. After mixing, the fermenting compost is again piled up to form a stack that kept for another 2 days. Polythene sheet covering is done for fermentation. After the latter two days of fermentation. The compost becomes ready for pasteurization.

(iii) Method 3—Wet cotton waste with approximately 65 per cent moisture is mixed with 5 per cent wheat bran and 2 per cent $CaCO_3$ and stacked to a pile of size 90 cm × 90 cm × 55 cm (h). The pile is covered with a poly sheet to ferment for 4 days with one turn in between. The compost is mixed with wet paddy in the ratio of 2:1. Now, the mixture is ready for disinfection by steaming in autoclave at 5 lbs for an hour or pasteurization as below (Kaur et al. 2004).

Pasteurization and cultivation methods: Pasteurization is done in the growing room made of polythene. In such case the mushroom house is made up on bamboo/wooden frame and covered inside with polythene sheet of 0.4 mm, while outside by polyform sheet of 1.27 cm, for insulation. The house is fitted with a ½ hp electric blower, a polyethylene air duct for ventilation and four windows, two on each long side of the house. Air vents are prepared near the upper corner of both gables. The ventilators are fixed with specially made ventilating fan to remove and introduce air from and to mushroom house, respectively, as and when required by simply operating the valves. The introduction of air inside the room is made through the multi perforated air duct (hole size 1 to 2 cm diameter) fitted under side the roof ridges. The shelves are prepared by bamboo and iron nets above 30 cm from ground keeping gaps 45–50 cm between two shelves. The compost is placed on the shelves up to a height of 10 cm. Boiler's outlet is fitted with a pipe which contains 2 to 4 cm diameter outlets at different places. The steam passes through the pipe to raise the temperature to 60°C for 2–4 hrs. Then the temperature lowered to 50–52°C by inserting fresh air through the specially designed ventilators. This temperature is maintained for next 8 hrs. The temperature is then lowered to 34–36°C gradually for 12–16 hrs and spawning is done during that time with 1.4 per cent (dry weight basis) spawn. The spawned bed is covered with plastic sheet for 3 days without giving any water and light. After three days, sprinkling of water and illumination of light with fluorescent tubes are given for fruit body formation in next 7–8 days.

Crop management

Watering and lighting: Paddy straw mushroom requires atmospheric temperature 30–35°C, the bed humidity 60–70 per cent, relative humidity 75–85 per cent and humidity in air around the bed 80–90 per cent for its growing as mushroom (Chang 1965, Yau and Chang 1972, Quimio et al. 1990). Further, it requires well lighted/illuminated condition for fruit body formation. Thus, mushroom house should be kept well illuminated by fluorescent tubes or cultivation should be done near window or at verandah. The bed is required to cover with a transparent polythene sheet keeping 6 inches gap between bed and sheet in order to maintain high humidity around the bed. Watering, as already discussed, is not required for the first 3–4 days, but thereafter regular sprinkling of water is needed to keep the mushroom bed moistened. However, at perhumid areas watering is not essential to get first flush during rainy season.

Harvesting and yield

Paddy straw mushroom should be harvested all at a time by twisting at base when a few of the mushroom fruit bodies of same flush reached at elongation or just opening stage. At that time, button, egg and elongation stages of mushrooms, which are most suitable stages for consumption, are available mostly. This harvesting stage remains at day time. The mushrooms of button and egg stages can also be exported after preserving in cans. Mushroom yields depend upon the cultivation seasons, variety of mushroom cultivated, methods of cultivation adopted and the stages of mushroom when that is harvested. In general, paddy straw mushroom is harvested at elongation and just opening stage for local use. However, sometimes mature fruit bodies are also harvested. As the size, shape and weight of different stages of mushrooms vary significantly, so the yield of paddy straw mushroom in traditional system may vary from 4.5 per cent to 28.3 per cent. However, on an average 1 kg of straw gives about 100 g of buttons and eggs of this mushroom, which means that the yield is 10 per cent of the substrate. The yield in improved method on cotton waste compost is about 40–45 per cent, while, in rice straw + cotton waste based compost is 25–30 per cent.

Packing and marketing

Paddy straw mushroom is highly perishable it cannot be kept even in refrigerator overnight, since it liquefies at 4°C. The fresh mushroom is marketed by packing in perforated hard paper box. The mushroom packing in perforated hard paper box can also be stored at 10–15°C for about 3 days (San Antonio and Fordyce 1972). However, the best measure to transport the paddy straw mushroom is after canning. The paddy straw mushrooms of button and egg stages are only suitable for canning. They are washed in water containing 0.1 per cent citric acid or 0.3 per cent sodium metabisulphite. Then, blanching is done by cooking the mushrooms at 95°–100°C for three to eight minutes in water containing 0.05–0.1 per cent citric acid. The blanched mushrooms are then put in cans (International standard: Size–16 oz of can for 454 g or 68 oz for 1930 g mushrooms) with citric acid or ascorbic acid @ 1 g/litre, leaving 1.5 cm of head space. The cans are then exhausted by passing over gas burners called steriflame for at least 3–8 minutes and hermetically sealed with the help of double seamier. The sealed cans are sterilized at 10 p.s.i. for 25–35 minutes or at 15 p.s.i. for 15 minutes. The cans are then cooled and stored in a cool and dry place.

Improvement in cultivation technologies

Deshpande and Tambane (1982) reported that spraying of 2 per cent glucose solution at button stage increased yield. Use of *Volvariella volvacea* strains, Vv S–4 of DMR, Solan and Vv–W of PAU, Ludhiana are found more productive (Kumari et al. 2008).

MILKY MUSHROOM CULTIVATION

The milky mushroom *Calocybe indica* (P and C), is also known as *dudh chhata* in West Bengal. It grows in summer and requires high temperature for its fruit body production. It is a long sized mushroom and grows in nature on humus under the road side trees or in forest. This mushroom has an attractive milky white sporophore with delicious flavour and longer shelf life. It is very suitable for the people suffering from hyperacidity and constipation due to presence of alkaline ash and high fibre content (Doshi et al. 1988). It is also suitable for pickling and chutney preparation. The cultivation method of this mushroom was first developed by Purkayastha and his coworkers (Purkayastha and Chandra 1976, Purkayastha and Nayak

1977, 1979, Purkayastha et al. 1981) and later, that was simplified successively by Chakravarty et al. (1981) and Trivedi et al. (1991, 1994). The production status of this kind of mushroom is still insignificant in the world.

Morphology and Other Characters

Sporophores of *Calocybe indica* grow solitary in soil, robust in size, centrally stipitate, fleshy, white or pale coloured, with to some extent pungent fungoid smell. Pileus is 10.0–14.0 cm in diameter, at first convex, latter expanded and flattened, white, non-hygrophanous, cuticle easily peeled, mat polished, sometimes appressed scales present at or around the centre, margin regular, incurved, smooth, nonstriate. Gills are distinctly formed, crowded, emarginated, separable, white, non-intervened, unequal, pliable, thick, attenuated towards margin, entire. Stipe is central, sometimes eccentric, cylindrical with sub-bulbous base, up to 10 cm long, white, cartilaginous, surface dry, fibrillose, base solid, without annulus and volva. Flesh is white. Hymenophoral trama are regular, slightly divergent below subhymenium. Basidia are with carmiphilic granules, clavate, tetrasterigmatic, 25.5–30.6 μm × 6.8–8.5 μm. Bosidiospores are hyaline, broadly ellipsoid, thin-walled, without ornamentation, with prominent apiculus, non-amyloid, 5.9–6.8 × 4.2–5.1 μm. Spore print is white (Purkayastha and Chandra 1985).

Facilities and Materials Required

Mushroom house or shed: Any type of house as used for dwelling purpose is suitable for milky mushroom cultivation. Thus, one can grow this mushroom in small scale at a corner of a house. However, for commercial cultivation, separate thatched/tinned house with bamboo mat wall plastered by cement and sand mixture, and cemented floor is well suited. The house is divided into two halves for spawn running and mushroom growing separately.

Substrate store house or shed: A small shed is required to store agricultural residues.

Water tank: Cemented tank with outlet or a plastic drum with outlet is required for soaking straw in water or in disinfectant solution for over night (16–18 hrs).

Water supply: Running water supply is needed to clean mushroom house, watering in mushroom bed and filling water tank.

Wooden, bamboo or steel racks: Wooden or steel rack of breadth 3 ft, height 8 ft, shelves 4, gap between two shelves 2 ft and length 18 ft, keeping gap 2 ft between the rack and wall and 3 ft between two racks with lowermost shelf 1 ft above the ground is used for cultivation of this mushroom.

Hand chopper/chaff cutter: One hand chopper or chaff cutter is required for cutting substrate materials.

Metallic drum: Required for water boiling when disinfection of substrate is done by hot water or vapour treatment.

Wire cage: Required if disinfection of substrate is to be done by vapour treatment.

Wooden plank: This is required for covering the bucket while hot water treatment is practiced in disinfecting substrate.

Plastic bucket: Required two plastic buckets for carrying substrate materials, spawn, etc.

Bamboo cane basket: Two bamboo cane baskets are required for removing excess water.

Sprayer: One sprayer is required for watering and chemical spray.

Gunny bags: These are required for hanging as wet in raising humidity in mushroom house during dry weather.

Fire wood or large gas oven: This is required for disinfection of substrate by hot water or vapour treatments.

Polypropylene bags: These are required for filling substrate mixed with spawn. The convenient size is 18″ × 14″.

Polythene sheet: This is required for spreading wet substrate in removing excess moisture.

Nylon thread: It is required for threading the mouth of poly bag.

Mosquito net: This is required for putting on windows and ventilators in preventing fly entry.

Substrates: Dried agricultural residues, most commonly paddy straw and wheat straw.

Spawn: Preparation of spawn is done with wheat grain or paddy grain. The methods used for paddy grain medium are already discussed in the chapter of spawn production. However, the standard method for this mushroom is with wheat grain. The wheat grain is boiled for 15–20 minutes then the excess water is decanted by a strainer. The surface of grains are dried in air and mixed thoroughly with calcium sulphate (gypsum) and calcium carbonate (chalk powder) @ 2 and 6 per cent (wet weight of grain), respectively. The spawn medium is now filled in a bottle or polypropylene packet and autoclaved at 15 lb pressure for 2 hours for sterilization. The sterilized medium is then inoculated with mushroom mycelia and incubated at 26–28ºC providing light for 6–7 hrs per day. During the course of spawn growth, the bottles are shaken on 6th, 9th and 12th day for uniform growth of mushroom mycelia. After three weeks of incubation, the spawns are ready for use.

Fungicide: Carbendazim (bavistin), mancozeb (indofil M-45) and zineb (indofil Z-78).

Antibiotics: Streptomycin sulphate and tetracycline hydrochloride mixture (plantamycin, pushamycin or agrimycin).

Insecticide: Malathion.

Disinfectant: Bleaching powder and formalin (for disinfection of room and rack after every crop).

Garden soil/loam soil: Required for preparing casing material.

Farm Yard Manure (FYM): Required for preparing casing material.

Sand: Required for preparing casing material.

Calcium carbonate: This is required for adjusting pH in casing soil.

Substrate Preparation

Substrate types

Fresh one year old paddy straw of *aman* paddy, wheat straw, pea haulms, tomato haulms, sugarcane bagasse, sunflower stalk, sawdust, bamboo leaves, coconut coir, etc. are more or less good substrates of this mushroom (Krishnamoorthy and Muthusamy 1997, Eswaran and Thomas 2003, Tandon et al. 2006, Biswas and Singh 2009a, Chaubey et al. 2010).

Processing of substrate

Substrates, like, paddy straw or other haulms of agricultural crops are cut into 5–8 cm long pieces with a hand chopper or chaff cutter. However, the crumpled paddy straw or soft haulms does not require any cutting.

Water soaking and disinfection

Water soaking and disinfection of substrates are done by various ways as convenient according to the available facilities. Here, two types of the methods which are commonly used and can be done by any farmers at their homes are described.

(a) Hot water treatment: In this method, substrate is water soaked for 16–18 hrs for overnight in a cemented tank or any large container. On next day morning, the substrate is taken out and kept on a cemented platform or in a bamboo cane basket in heap for about 1 hr to remove the excess water from the substrate. After one hour of water removing, the water soaked substrate is kept in a large vessel/bucket and there boiled water is poured so that the substrate is fully immerged in hot water. The whole set is covered by a wooden plank and kept for 2 hrs. Time to time boil water may be added to maintain the temperature of above 65°C. The substrate is then taken out in a bamboo cane basket for draining out the excess water and later spread in sun to remove excess moisture for about 1–2 hrs depending upon the brightness of sunlight. The moisture is kept 60–70 per cent in substrate and that can be visibly identified by seeing the dryness of the surface layer of substrate. In rainy season, removal of moisture may be done by spreading under roofed/covered condition for longer period, even if it required 1–2 days that should be allowed, since high moisture will spoil the substrate in polythene bag during spawn running stage. Air circulation can be raised by electric fan or other device, if possible.

Alternatively, in other way, the chopped straw is soaked in water for a certain period and boiled in a container for about 30 minutes. Then, the excess water is drained out and cooled the straws by spreading on a clean surface before using that for mushroom bed preparation.

(b) Steam pasteurization: Steam pasteurization is normally done in a peak heating room or in bulk chamber both facilitated with boiler and blower as discussed in the preceding chapter. However, alternatively for simplification, a top open metallic drum is placed on a brick made oven and filled in water up to 1/5th portion by pouring two buckets of water. The water is then boiled by burning fire wood or gas at bottom. The overnight soaked substrate after removing water is put in a wire net cage which has four stands and placed inside the drum as such that the water level remains below the bottom of the cage. The drum is then covered by a lid with a small opening at centre or by water soaked two gunny bags placing cross wise. The steam coming out from the boiling water passes through the substrates and finally through the opening of lid or leakages of gunny bags. After, 45–60 minutes of steaming, the cage along with substrate is taken out and the substrate is sprayed out in sun for removing excess water.

Note: *Although, chemical disinfection is practiced in certain cases, it is not advocated for disinfecting the substrates of milky mushroom. Since, the chemicals retard the growth of mushroom and the best result is observed when the substrate is pasteurized in steam followed by disinfection in hot water (Sharma et al. 2006).*

Cultivation

The milky mushroom requires 25–35°C for its growth. It grows on a wide range of agricultural wastes and like oyster mushroom its conversion rate is very high (Biswas and Singh 2008). It is very easy for cultivation and can be grown in different methods of cultivation.

1. Preparation of bag and casing soil

Method 1:

(i) Bag filling—The process of milky mushroom cultivation is same as oyster mushroom cultivation in polythene bag up to the stage of spawn running. Mushroom beds are prepared in various ways, but the most convenient way is that in which bed is prepared in polythene bag. In this method, transparent polythene bag (18″ × 14″) is perforated (hole size 0.4″ diameter) at a distance of 4″ all over its body surface (Figure 7.6). Disinfected paddy straw

Figure 7.6 Milky mushroom cultivation on floor and rack.

(about 0.75 kg on dry weight basis) as made ready after any of the earlier methods of disinfection and a half bottle/bag of spawn (75 g) are used to prepare a bed (Biswas and Singh 2008b). The substrate is placed inside the bag in 4–5 layers by pressing with hand palm. After completion of one layer, 10–15 g of spawn grains are spread over the substrate layer. Then, another layer of substrate is prepared and the spawn grains sprayed over that similarly. In this way 4/5th portion of the transparent polythene bag is filled up by the spawned substrate. Now, the bag is closed tightly by tying with a piece of nylon thread and incubated in spawn running room in dark for about 18 days.

(ii) Casing material—During spawn running stage (at least 6 days earlier of opening of bag), casing soil is made ready by mixing field soil, sand and well decomposed FYM in 1:1:1 ratio. The mixture is then suspended in boiled water for about one hour. After completion, the excess water is removed by decanting the whole material on a clean surface. The material is kept as such for drying enough in order to make it fragile by hand palm pressure.

(iii) Casing—When the spawn running is complete in polythene bag, the mouth of the bag is unthread or cut open and the casing soil as prepared earlier is spread over the spawn run substrate up to 1″ height of thickness. Alternatively, the fully grown bag may be cut into two halves and the cut surface of each halve is cased.

(iv) Fruit body production—The mushroom bed after casing is kept on a rack or on a raised floor made by brick for allowing fruit body production repeatedly from the upper surface. Generally it takes 40–45 days for harvesting fruit body from the date of inoculation of spawn.

Method 2:

(i) Bag filling—Straw is chopped into small pieces (2–4 cm size) and filled in a gunny bag. The gunny bag with straw pieces is dipped in a water tank for soaking water for about 12–16 hrs. The excess water from straw is decanted and the wet straw is pasteurized by steam at 65°C for 5–6 hrs or by hot water treatment at 80–90°C for 40 minutes. Polythene bag filling is done as discussed in earlier method with the use of 4–5 per cent spawn of wet weight of substrate. The bags are now incubated for 20 days at room temperature for spawn running. After completion of spawn run the bag may be cut into two halves or open the mouth. The cut or open surface is pressed with hand palm and casing material is applied up to 2.5–3 cm thickness.

(ii) Casing soil formulations—In most cases casing soil is prepared by mixing 75 per cent loam soil and 25 per cent sand. The pH of the soil is adjusted to 7.8–7.9 with the use of calcium carbonate (generally 10 per cent is used). The pH adjusted soil is now sterilized in autoclave at 15 lb pressure for an hour.

Alternatively, casing material can also be prepared by mixing 2-year old spent compost, FYM and sandy soil in 2:1:1 ratio and treating with formalin (4 per cent), covering that with polythene sheet for 72 hours (Tandon et al. 2006, Sukla 2008, Biswas and Singh 2009a, Chaubey et al. 2010). Use of clay loam soil and 2-year old FYM in 1:1 ratio; 2-year old FYM and sand mixture in 1:1 ratio; 2-year old FYM and spent compost in 2:1 ratio; 2-year old FYM and coir pith in 2:1 ratio; 2-year old spent compost, FYM, sand and garden soil in 1:1:1:1 ratio or the mixture 2-year old spent compost and FYM in 2:1 (differently1:1) ratio as casing material is also found more productive in different studies (Singh et al. 2007, Sukla 2007, Bhatt et al. 2007, Dayaram 2009).

(iii) Casing—Casing is done on the top surface of mushroom bed after opening the mouth. The casing material is spread uniformly up to a height of 2–3 cm. Sometimes, the polyethylene bag with myceliated substrate is cut into two halves. The cut surfaces are then pressed with hand palm and casing is done on the cut surfaces. By this way two units of mushroom production are made from a single bag.
(iv) Fruit body production—The mushroom bed is now kept on a rack for allowing fruit body production repeatedly from the upper surface. Generally it takes 40–45 days for harvesting fruit body from the date of inoculation of spawn into the substrate in polythene bag.

2. Watering and lighting

The mushroom requires well illuminated condition for long time along with high relative humidity (80–95 per cent) for its fruit body development. So, the room should be fixed with fluorescent light or cultivation should be done near window side. To raise relative humidity during summer wet gunny cloth may be hanged at different places inside the room. The watering to the bed, i.e. on casing layer, is to be done regularly. Care should also be taken that mushroom bed (bag) is not over watered. So, a few holes at the bottom of the bag are prepared during the time of casing to run off the excess water, if any.

3. Harvesting and yield

Mushroom fruit bodies are to be harvested before fully opening of the cap, i.e. when the margins are still rolled downwards. The harvesting can be advanced or delayed one or two days depending upon the marketing and consumption facilities, since it takes quite long time for spoiling in beds. The fruit body is to be hold by fingers and twisted to remove the mushroom from the substrate by gentle pulling. The soil and straw adhered base of mushroom is to be cleaned by a knife. The mushroom yields about 1 kg from 1 kg paddy straw substrate (i.e. BE-100 per cent). Supplementation of ground soybean meal @ 4 per cent in the paddy straw substrate enhances the mushroom yield further (Singh et al. 2007).

4. Packing and marketing

As the shelf life of the mushroom is quite long period so it can be brought to market as open as applicable for other vegetables. However, packing in perforated hard paper box and maintaining 10–15ºC during transportation or storage is advisable.

SHIITAKE MUSHROOM CULTIVATION

Shiitake mushroom (*Lentinula edodes* = *Lentinus edodes*) is named after the host tree 'Shii' (*Castanopsis cuspidate*) wherefrom this mushroom was originally collected in Japan (Ito 1978). In China this mushroom is referred as 'Xiang-gu' (literally rice smelling). Nowadays, this is the most popular mushroom in east Asian countries, particularly in China, South Korea, Taiwan and Japan where its production in 2003 was 2,228,000 mt (21.8 per cent of total mushroom production of China), 41,876 mt (24.6 per cent of South Korea), 36,000 mt (33.4 per cent of Taiwan) and 35,294 mt (10.7 per cent of Japan), respectively (Chang 2007). As per the global mushroom production it ranks second. This mushroom has been grown for many years on thin wooden logs in Asia. It has a unique taste and flavour with enormous medicinal values (Chang and Miles 1987, Hobbs 1995, Mizuno 1999, Stamets 1999). The water soluble polysaccharide of high molecular weight chemical, lentinan, present in the mushroom reduces plasma cholesterol in human and controls blood pressure (Doshi and Sharma 1995, Wasser and Weis 1999). It also has anti-cancer, anti-HIV, anti-diabetic properties; it acts as anti-measles in children and also has ability of boosting immune system (Chihara 1992, Jong and Brimingham 1993 and Mizuno 1999, Wasser and Weis 1999). Further, the cell free extracts and exudates of this mushroom exhibit antimicrobial activities (Hirasawa et al. 1999, Dewangan et al. 2007). In nature it grows on dead parts of hard wood trees, like, *Quercus* sp.

Morphological and Other Characters

Sporophores usually grow on wood of dead deciduous trees of Fagales, e.g. oak (*Quercus* sp.), chestnut (*Castanea* sp.), horn bean (*Caprinus* sp.), beach (*Fagus* sp.), etc. They are usually eccentric, but sometimes centrally stipitate, arising solitary or in small groups. Pileus is pale to dark reddish brown, deep brown to dark tan towards centre, dry and more less fibrillose, the cuticle breaking into scales of various sizes and shapes, often rimose with white context showing more frequently along the periphery, up to 11 cm broad, at first convex latter depressed, margin regular with persistent veil remnants. Gills are white, becoming spotted reddish on injury, gradually flushed browning on aging, crowded, edge denticulate, not free from stipe, trama with thick walled interwoven hyphae, hyphae hyaline. Stipe is usually 3.0–5.0 × 0.8–1.3 cm, solid, rod-shaped, surface rough with brownish scales, pale reddish brown annulus and volva are absent. Flesh is white or brownish near the cuticle, firm and tough-fleshy in age, softer in immature, taste acidulous, odour slight, but not distinctive. Hymenophoral trama are regular. Basidia are normally tetrasterigmatic with four basidiospores, but sometimes basidium contains five or even six spores. Basidiospores are cylindrical to elliptical, non-amyloid, smooth, thin walled, hilum present. Pleurocystidia are absent. Hyphal system is monomitic, scarified generative hyphae are present, skeletal or ligative hyphae are absent, generating hyphae loosely interwoven, inflating, constricted septa with clamp connections (Nakai and Ushiyama 1974, Smith 1978, Tokimoto and Komatsu 1978, Purkayastha and Chandra 1985).

Facilities and Materials Required

Wooden Logs: Wooden logs of *Quercus* sp. are generally used for log cultivation.

Sawdust: Sawdust of non-resinous hard and soft wood is required for poly bag cultivation.

Autoclave (large): This is essential for sterilizing the substrate mixture (sawdust and rice bran or wheat bran) required for mushroom growing.

Growing room: Environmentally controlled growing room. It grows below 20°C.

Ice: Ice is required for giving chilling treatment for fruit body formation.

Spawn Preparation

Sawdust spawn

Mixture of sawdust and rice bran (4:1 ratio) are water soaked with required amount of water and autoclaved, inoculated with mycelium and incubated at temperature between 24–28°C for

3–4 weeks (Ito 1978). Sometimes, wood pieces of wedge or small cylindrical shaped are used instead of sawdust to prepare wood plug spawn. Later one is most frequently used for log cultivation. Different formulations of the ingredients are used for preparation of saw dust spawn (Suman and Sharma 2005). These are as follows:

Formula 1: The mixture of sawdust (65 per cent), wheat bran (25 per cent), used tree leaves (20 per cent) and water (65 per cent).

Formula 2: The mixture of sawdust (75 per cent), wheat bran (20 per cent), sucrose (1 per cent), calcium carbonate (1 per cent) and water (65 per cent).

Formula 3: The mixture of sawdust (800 g), rice bran (200 g), sucrose (30 g), calcium carbonate (6 g), potassium nitrate (4 g) and water (2 litres).

Grain spawn

The boiled wheat or sorghum grains are mixed with 10 per cent sawdust and 0.5 per cent wheat bran. This is filled in spawn packets (polypropylene bags) and sterilized in autoclave at 20 lbs pressure for about 2 hrs (Doshi and Sharma 1995).

Cultivation

The four important strains, viz. Le 4/03 (Korea), LeOE 9 (DMR, Solan), LeOE 142 (DMR, Solan) and Le 7/04 (China), and their hybrids, like LeOE 9 × LeOE 142, LeOE 9 × Le 7/04, Le 4/03 × LeOE 9 and Le 4/03 × Le 7/04 are generally used/identified for cultivation in India (Samota and Sharma 2008).

Log cultivation

This mushroom is generally cultivated on moderately thin wood logs of dead deciduous trees, like oaks (*Quercus serrata*, *Q. acutissima*, *Q. mongolica*), Shii (*Castanopsis cuspidata*), chestnut (*Castanea crenata*), hornbeams (*Carpinus* sp.), etc. Of the tree species, *Quercus serrata and Q. acutissima* are most suitable (Ito 1978). The dead wooden logs (1–1.5 m length, 5–15 cm diameter) are used for mushroom cultivation. At 20–30 cm apart on the long axis of the wood log, small holes of 1 × 1 cm in size and 1.5 to 2 cm in deep are prepared. In case of thick logs, several rows of holes are prepared keeping 6 cm distance between two rows. Sawdust spawn or wooden plug spawn (not discussed) is filled in the holes and covered the filled up holes with hot paraffin wax. The inoculated logs are then kept in 'laying yard' in evergreen groves or shady places in a flat pile for development of the mycelium. The pile is covered by straw or by gunny bags. Time to time light watering is needed for keeping moist condition. However, it should be remembered that excessive moisture causes contamination. This condition is kept for 8–12 months. The optimum temperature for mycelial growth is 24–28°C, so care should be taken to maintain this temperature. After completion of mycelial growth, the logs are transferred to the 'raising room' where the temperature remains 12–20°C. The raising rooms are polythene top covered shade with surrounding wall of black vinyl netting to minimize light intensity and facilitate ventilation. The logs are kept in lean condition on a bamboo support and sprayed with chilled/cold water or immerged in cold water for 1–2 days then kept in lean condition for initiation of fruit body formation. Relative humidity in the raising room is maintained around 80–90 per cent. Primordia develop within 24 hrs of cold water spraying and thereafter within 7–14 days fruit bodies become ready for harvest. Generally, three flushes are harvested in a year keeping gaps of 30 days. After completion of harvest in a given year, the logs are shifted to the laying yard and kept in favourable condition as done earlier. In this method, the same log gives production for 3–6 years (Ito 1978, Doshi and Sharma 1995, Suman and Sharma 2005).

Bag cultivation

In case of bag cultivation, hardwood sawdust of oaks (*Quercus* sp.) or maple (*Acer* sp.) is used. This sawdust is mixed with rice bran or wheat bran for favouring mycelial growth. The pH of the substrate mixture is adjusted to 5.5–7.0 with the use of required quantity of calcium carbonate ($CaCO_3$) or sodium carbonate (Na_2CO_3). Different formulae are used to make the substrate mixture. In most cases saw dust is the major ingredient. Apart from saw dust certain agricultural residues are also used for making substrate mixture. These are as follows:

Formula 1: Sawdust (80 per cent), rice bran (20 per cent) and water (65 per cent).

Formula 2: Hardwood sawdust (89.8 per cent), rice bran (10 per cent), $CaCO_3$ (0.2 per cent) and water (60 per cent) (Han et al. 1981).

Formula 3: Sawdust (80 per cent), millet (10 per cent), wheat bran (10 per cent) and water (60 per cent) (Royse et al. 1985).

Formula 4: Hardwood sawdust (100 kg), rice bran (16 kg), corn powder, rice or millet (8 kg), $CaCO_3$ (1.0 kg) and water (62 per cent).

Formula 5: Sugarcane begasse (50 kg), rice bran (10 kg), gypsum (1.5 kg), potassium sulphate (15 g), urea (15 g), magnesium sulphate (10 g) and water 60–65 per cent of the substrate mixture.

Formula 6: Corn cobs (40 kg), sawdust (10 kg), wheat bran (12.5 kg), cane sugar (1 kg), pectin (15 g), urea (20 g) and water 60–65 per cent of the substrate mixture.

Formula 7: Hardwood sawdust of broad leave plants (40 kg), wheat bran (8 kg) and water (65 per cent) (Ahlawat and Verma 2000–2001).

Formula 8: Sawdust (80 per cent), wheat bran (20 per cent) and water (65 per cent) (Meena and Sharma 2005).

Formula 9: Sawdust (80 per cent), wheat bran (20 per cent), calcium carbonate (1 per cent) and moisture (65 per cent) (most frequently used, Samota and Sharma 2008).

(a) Method 1—Substrate mixture following any one of the formulae may be used for cultivation. Sawdusts are soaked in water for about 18 hrs and added other ingredients as per formula. These ingredients are mixed thoroughly and filled in polypropylene bags @ 1.5–4 kg/bag and a stick/spike is introduced in the centre for making hole. Now, the bags are sterilized in autoclave at 15 lbs for 1/2 hr or by steaming. Certain amount of pressure with hand palm is needed to make the substrate mixture in compact cylindrical form. Inoculation is done aseptically with saw dust spawn at the central hole (about 15 mm diameter) which appeared after taking out the stick or spike. The spawning is done with sawdust spawn (10–15 g). The spawn is inserted in the hole with the stick/spike which is taken out earlier. The spawned hole is covered with 33 mm square medical adhesive tape. The inoculated bags are kept in stack in growing room for 2 months at 24–28°C.

Alternatively, the substrate filled polypropylene bags may be closed with cotton plug supported by iron or heat stable plastic ring. These are sterilized in autoclave at 15 lbs for 1/2 hr and inoculated with spawn @ 2 per cent of wet weight of substrate aseptically by opening cotton plug. Thereafter, they are kept in stacks in growing room at 24°C.

During spawn run bags are kept under low light intensity and high humid place with little ventilation. Water spraying is not required during this stage. If any contamination appeared in bag, that is discarded. After 2–4 weeks the mycelia form a thick sheet on the substrate. Bumps of mycelia due to clumps are formed at several places on the mycelial sheet. At this stage a little aeration is to be provided. The spawn running stage completes in about 60 days. After completion of spawn run, the bags are kept as it is for browning of the mycelia. Now, the polypropylene bags are removed and the myceliated brown substrate block is dipped in chilled water (4–5°C) for about 5 minutes. This cold shocking is to be repeated for 2–3 days. This period is known as 'induction'. Then, the block is kept in single layer at temperature below 20°C (preferably at 12–18°C) in a growing room for fruiting. The light intensity 500–1000 lux, well ventilation for aeration and relative humidity 85–95 per cent are required for fruit body formation, although, for the growth of fruit body the humidity of 60–80 per cent is optimum. Fruit body develops within 7–14 days after cold shock (Figure 7.7A). After

A

B

C

Figure 7.7 *(Contd.)*

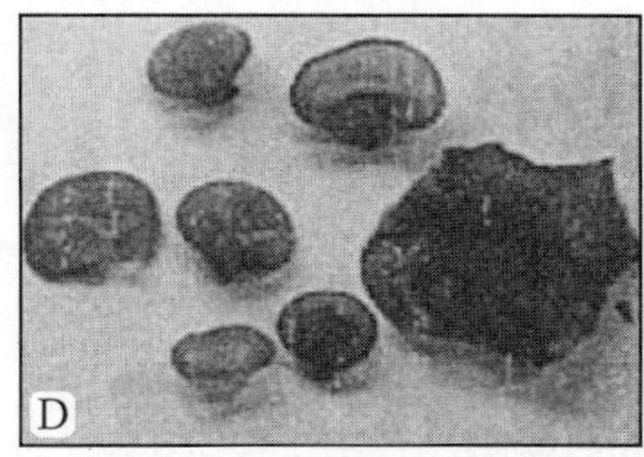

Figure 7.7 Specialty mushrooms. **A.** Shiitake mushroom, **B.** and **C.** Parasol mushroom, **D.** and **E.** Ear mushroom.

completion of harvest, the block is again transferred in growing room at 20–30°C for mycelial growth. This period requires about 7–21 days. Thereafter cold shock as done earlier is given for fruit body formation. Later, it is incubated at 12–18°C for fruit body development. By this way, two to three flushes can be taken in a year.

(b) Method 2—In another method, sawdust is soaked for 24 hrs and then removed water by decanting. This wet sawdust is now spread on a polythene sheet and exposed to sun for about 1–2 hrs to remove excess moisture. Now, the other ingredients, like, wheat bran (20 per cent) and calcium carbonate (1.0 per cent) are mixed. The substrate mixture is filled in polypropylene bags (16″ × 12″ of size). The bags are closed with cotton plug supported by iron or heat stable plastic ring. This is sterilized in autoclave at 15 lbs for 1/2 hr and inoculated one day later with spawn @ 2 per cent of wet weight of substrate aseptically by opening cotton plug. Thereafter, it is kept in a stack in growing room at 22–24°C for spawn run. After completion of spawn run, the plastic covers are removed and the bed logs (the myceliated sawdust-substrate mixture) are kept as open for about 4–5 weeks to an environmental conductive for browning of the exterior log surface. Frequent watering is done on substrate logs during that time and maintained room humidity 80–90 per cent. After browning, the logs are dipped into water (5–10°C) for 1–2 hrs in chilling apparatus and kept on racks each day until blister and bumps started initiating. The bumps develop into pin heads which within 72 hrs form mature fruit bodies (Samota and Sharma 2008).

Harvesting

Harvesting of fruit bodies is done at an early stage of development by hand plucking. About 200–250 g of fruit bodies may be harvested per kg of dry substrate. Sharma et al. (2006) observed higher yield with BE of 62 per cent on sawdust substrate, while, compared with wheat straw (BE 34 per cent) and other substrates.

Note: *As per the physiological requirement it is evident that malt extract agar medium with pH 5–8, temperature 20–25°C and regular cold shock for 60 minutes at 5°C are suitable conditions for the mycelial growth of the mushroom (Samota and Sharma 2006).*

BLACK EAR MUSHROOM CULTIVATION

The black ear mushroom also known as Jew's ear, Wood ear or *Mou-Erh* (in Chinese) belongs to the genus *Auricularia* order Auriculariales. It is characterized by gelatinous, irregular and foliose fruit bodies [Figure 7.7(D) and (E)]. Although all the species of genus *Auricularia,* found in nature are edible, only two, viz. *Auricularia auricula* and *A. polytrich,* are considered as popular edible species of this category (Cheng and Tu 1978). The former one is generally collected from the local habitats and the latter one is cultivated. In nature, they grow on wood logs. As per the global position, this mushroom ranks 4th with production of 485,000 mt/year, which shared 7.9 per cent of total mushroom production in 1997 (Chang 1999). This mushroom has great demand in China, Japan and Philippines. Thiland is the major importer of this mushroom for local use. About 80 per cent of the dried mushroom produced in Taiwan is exported to Hong Kong, Japan and USA (Garasiya et al. 2007).

Morphology and Anatomy

Auricularia auricula

Sporophore is tough gelatinous, sessile, sometimes with short stipe, up to 12 cm broad, caespitose, foliaceous, ear shaped, yellow brown to reddish brown in fresh, yellow brown to olive brown when dry. There are several distinct hyphal zones on transverse section of the fruit bodies as follows: *Zona pilosa,* this is the outer most layer consisting of hyaline hairs of 80–100 μm long and 5.0–7.0 μm broad (dia.), hyaline, without central strands, tips rounded not in dense tufts. *Zona compacta* is 65–75 μm thick. *Zona Subcompacta superior* is 115–130 μm thick. *Zona intermedia* is 280–300 μm thick. *Zona subcontacta inferior* is about 100–200 μm. Hymenium is about 150 μm thick, red brown when moist, dark when dry, smooth. Basidia are cylindrical, 50–60 × 5–6 μm. Basidiospores are 10–15 × 5–6 μm, allantoid (Cheng and Tu 1978).

Auricularia polytricha

Sporophore grows on dead wood, leathery or rubbery gelatinous when fresh, brittle or cartilaginous on drying, ear shaped, i.e. cupulate, sessile sometimes with short stalk, up to 10 cm broad, red brown when fresh, grey or tan on drying. There are eight distinct hyphal zones on transverse section of the fruit bodies: *Zona pilosa* is the outer most layer, consisting of hyaline hairs of 400–500 μm long and 5.0–7.0 μm broad, arranged in dense tufts with prominent central strand. *Zona compacta* is 20–25 μm thick, compact indistinguishable hyphal elements. *Zona subcompacta superior* is 75–90 μm thick, hyphae arranged perpendicular to the surface. *Zona laxa superior* is 250–260 μm thick. Medula is 25–30 μm thick. *Zona laxa inferior* is 250–260 μm thick. *Zona subcompacta inferior* is 90–100 μm thick. Hymenium is 80–90 μm thick, dark, smooth and occasionally papillate, basidia are cylindrical, 50–60 × 4–6 μm. Basidiospores are 12–17 × 5–6 μm, allantoid (Cheng and Tu 1978, Purkayastha and Chandra 1985).

Facilities and Materials Required

The materials and facilities as required for oyster mushroom cultivation are enough for cultivating ear mushroom. The following things are required mostly:

- (i) Wheat or paddy straw
- (ii) Wheat bran
- (iii) Wooden logs (for log cultivation)

(iv) Hot water or steam or autoclave
(v) Polypropylene bag

Spawn

Two types of spawns are generally used for the cultivation of mushrooms.

1. Grain spawn

Grains of wheat or sorghum are used. The spawn preparation is same as done in common mushroom cultivation with the mixture of calcium carbonate (1 per cent) and calcium sulphate (3 per cent).

2. Sawdust spawn

The spawn is prepared in the medium consisting of sawdust (100 g), rice bran (25 g), potassium nitrate (5 g), calcium carbonate (7 g) and water (65 per cent). The inoculated spawn becomes ready within 3–4 weeks of incubation at 25–26°C.

Several other formulae are also used alternatively for preparation of saw dust spawn. These are as follows: (a) Sawdust 65 per cent, wheat bran 15 per cent, used tea leaves 20 per cent and water 65 per cent of the mixture, (b) Sawdust 78 per cent, wheat bran 20 per cent, sucrose 1 per cent, calcium carbonate 1 per cent and water 65 per cent of the mixture, and (c) Sawdust 880 g, rice bran 320 g, sucrose 30 g, potassium nitrate 4 g, calcium carbonate 6 g and water 2 litres. All the substrate ingredients as per a given formula are thoroughly mixed and soaked with required amount of water. They are filled in poly propylene bags, plugged with nonabsorbent cotton and sterilized in an autoclave at 20 lbs p.s.i. for 2 hrs. The inoculation with mycelia is done aseptically after cooling. The inoculated spawn packets are incubated at 25–26°C for 30 days.

Note: *Between the two types of spawn, sawdust spawn is better.*

Cultivation

Bag cultivation

The wheat or paddy straw is water soaked for over night. Then excess water is removed. This wet straw is mixed with wheat bran (4–8 per cent, better 10 per cent) and filled in polypropylene packets closed with cotton plug. The filled up bags are sterilized in an autoclave at 15 lbs pressure for 2 hrs. Alternatively, boil water treatment or steam disinfection can be made for substrate disinfection. However, the former method is mostly used. After cooling, the bags are inoculated aseptically with spawn @ 2–4 per cent (better 4 per cent) on wet weight basis. The inoculated bags are incubated in spawn running room at 25°C. After 2–3 weeks of incubation when the spawn run is completed; the bags are cut longitudinally with a blade at two to three places to give longitudinal slits. Now, the bags are kept at 22–26°C in high humid (R.H. 85 per cent) place of growing room where enough light and fresh air are present. In certain cases, forced aeration by air handling unit and illuminated light are given for about 2–3 hrs/day. Water is sprinkled twice a day for appearance of fruit bodies. The pin heads appear after 23–26 days of slitting and for maturation of fruit body after pin head stage 3–7 days are required. Thus, the fruit bodies become ready for harvest after 30 days of slitting the bag (Bhandal and Mehta 1989, Doshi and Sharma 1996, Ahlawat and Verma 2000–2001, Garasiya et al. 2007, Rawal and Sharma 2010).

Log cultivation

Wooden logs of *Morus* sp. *Ficus* sp. *Sesbania* sp. *Accacia* sp. *Betula* sp., *Fagus* sp. and many other broad trees are used for this mushroom cultivation. Inoculation is done within 7 days of the logs are felled, since; the water content in log should remain 80 per cent. The logs with approximately 3–6 cm diameter are cut into 1 metre in length. Small holes 1 cm × 1 cm and 1.5–2.0 cm deep are made with drilling machine or chisel and hammer. The holes are made at a distance of 20–30 cm apart in long axis, keeping gap of 6 cm between rows. The holes are now filled with sawdust spawn and covered with the bark and melted paraffin wax. The inoculated logs are placed as tilted condition on a log placed on the ground as such that these make a very small angle to the ground. This condition favours mycelial growth. The temperature of the laying yard is kept 20–33°C. In about 30–40 days of inoculation the mycelial growth is completed. Now, the logs are placed in raising room in tilted condition on some bamboo supports. Sprinkling of water is done regularly. The temperature is maintained 23–28°C in raising (cropping) room. The humidity inside the room is kept about 85 per cent. Suitable natural light or illuminated light is allowed for fruit body formation. The fruit bodies develop from the inoculated logs for about 5–6 months in a year. The same log gives fruiting for 10–12 years. In case of dryness appeared in log, the log is then immerged in water for a day.

Harvesting

Harvesting is done by hand plucking. The fruit bodies are harvested after 7–8 days of primordial formation. Biological efficiency of about 80–130 per cent is observed for wheat straw substrate supplemented with 4 per cent rice bran (Bhandal and Mehta 1989). The wheat straw supplemented with 10 per cent wheat bran gives more fruit bodies with BE 93 per cent Rawal and Sharma 2010). One piece of Sesbania log (2–15 cm diameter) produces 3–5 kg mushrooms in 5–6 months. The fruit bodies are stored in dry condition after sun drying.

SILVER EAR MUSHROOM CULTIVATION

The silver ear mushroom (*Tremella fuciformis*) also known as 'white jelly fungus' belongs to the family Tremellaceae, order Tremellales and class Basidiomycetes. It is a very precious mushroom in South East Asian Countries, since, it has significant medical values in curing several diseases, like tuberculosis, blood hypertension, common cold, etc. It also gives nutrition required for skin care and improves personal appearance. Further, it increases the vigour and extends life span of human beings (Lee 1596, Chen and Hou 1978). As per the world production, it ranks 7th with production of 130,000 mt/year (Chang 1999, based on 1997 data). Its production is mainly confined in the East Asian countries, particularly, in China and Taiwan (Chen and Hou 1978, Tu 1981).

Morphology and Other Characters

The silver ear mushroom forms firm gelatinous translucent to opaque white basidiocarps with no strange flavour. It grows saprophitically on woods of various plant species. The basidiocarp is blade shaped, lobed and curved, 5–15 × 4–12 cm in size, 0.5–0.6 cm in thickness, agar like at first, becoming soft and mucilaginous, often acquiring a sordid tint with age. Hymenium spreads over whole surface of the basidiocarp. The base is spread out into a yellow cartilaginous sheet beneath the bark of wood. On drying, the basidiocarps remain as yellowish horny layer that again revive in humid condition. Mycelia on agar medium are hyaline, with clump connections. Basidia are obovate, 6–11.4 × 7.2 μm, closely packed, longitudinally divided into four cells. Epibasidia are 12–16 × 2.5–3 μm in size, with one basidiospore. Basidiospore is subglobose to obovate, granulate, 4–7.5 × 6.8–11.6 μm in size, slightly attenuated towards one end (Chen and Hou 1978).

Facilities and Materials Required

Wood logs: Thin wood logs (1–1.2 m long) of mango (*Mangifera indica*) and other broad leaf trees are used for silver ear mushroom cultivation. The wood logs after two to three months of felling of trees are used for mushroom cultivation.
Raising shed: Thatched shed or plastic covered shed is generally used as raising shed.
Bamboo: Bamboo structure is needed as support to tilt the log for raising fruit bodies.
Water: Water in pipelines is used for spraying on logs.

Spawn

Spawn of silver ear mushroom is prepared in the medium consisting of a mixture of sawdust (80 per cent) and rice bran or millet grains (20 per cent). This mixture is added with water 60–65 per cent of the mixture, stuffed in bottles or polypropylene bags and sterilized at 20 lbs p.s.i. for an hour (Chen and Hou 1978, Tu 1981). The sterilized spawn medium is inoculated aseptically with spores or a bit of tissue of fruit body or mycelia grown in culture medium. Incubation is done at 23–26°C for about 2 weeks for growing mycelia in the spawn medium. Thereafter, the spawn with fully grown mycelia is kept at 15°C for another 2–3 weeks before use (Chen and Hou 1978, Tu 1981).

Cultivation

The mushroom grows well in the temperature range between 20°C and 28°C with relative humidity 88–95 per cent, although, its vegetative growth occurs optimum at 25°C (Chen and Hou 1978). Wooden logs (1–1.2 m long) of mango (*Mangifera indica*) or other broad leaved trees are drilled or punched to make bed logs and small amounts of spawn are inserted in the holes. The outside openings are closed with bark and paraffin. The size of holes is kept 1–1.5 cm in length and breadth with a depth of 1.5–3 cm, depending on the size of logs. The number of holes per log corresponds to the outer surface of the logs. In one metre log the number of holes is 3.5 times of its diameter measuring in cm and in 1.2 m log the number of holes is 4.5 fold of the log diameter (cm). After inoculation, the bed-logs are placed at a suitable place, i.e. in laying yard, for mycelial development at 20–25°C. In commercial cultivation, laying is done in raising room. The logs at laying stage are placed obliquely in upright position on a single log laid cross wise, and thus, the other logs made a small angle with the surface of the ground Figure 7.8. This condition is kept for 35–45 days. At this stage no excessive moisture is needed, and thus, the logs are sprayed with little or no water. When the mycelia developed thoroughly on the wood, the logs are transferred to raising room (thatched or polythene covered shed) and kept in a single layer on a bamboo support in more upright condition. This operation is termed as 'raising' and it needs frequent watering to keep the bed-logs as moist and the relative humidity at 85–95 per cent in the raising room. The temperature 20–27°C is maintained inside the raising room throughout the growing period. Fruit bodies begin to develop after about two months of inoculation. The fruit bodies develop first from the inoculated sites later from other sites of the logs. Fruiting stage continues for about seven months. The same bed-logs are used for growing mushrooms for several years by keeping the logs in laying yards after immerging in water for a day.

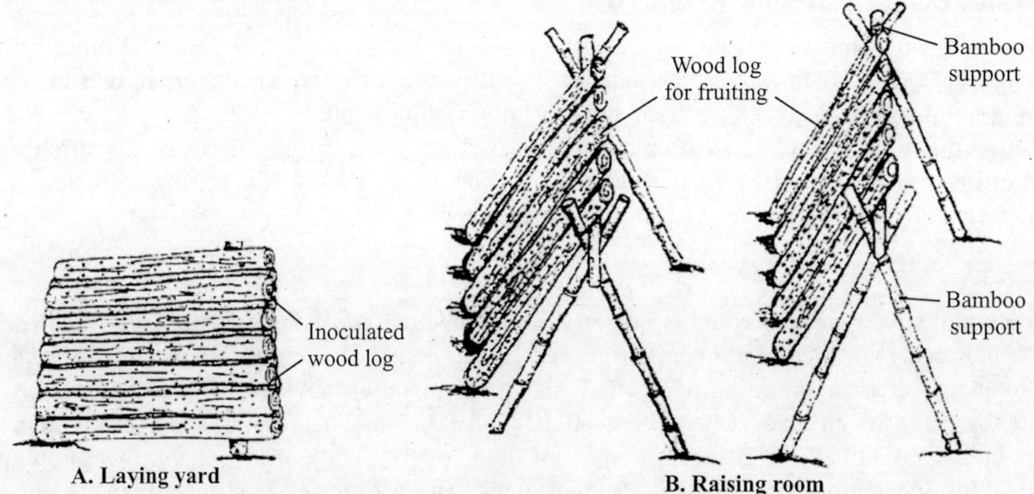

Figure 7.8 Diagrammatic view of silver ear mushroom cultivation. **A.** Arrangement of wood logs in laying yard, **B.** Arrangement of wood logs in raising room.

Harvesting and Packing

Fruit bodies of about 10–15 cm in diameter are harvested by hand plucking. They are washed with clean water to remove dust particles. The fresh mushrooms are packed in perforated paper boxes to use as fresh for food. It does not change any colour during transportation and short-term storing in markets. The mushrooms are generally dried in oven for about 8 hrs, initially at 50°C and that later gradually changed to 40°C. They are packed in plastic bags and stored in airtight boxes.

WINTER MUSHROOM CULTIVATION

The winter mushroom *Flammulina velutipes,* belongs to the family Tricholomataceae of order Agaricales. This mushroom is also known as by various other names, like Golden needle mushroom, Enoki, Enokitake (in Japan), velvet foot mushroom and velvet stem mushroom, at different places. In nature it grows during winter as wood decaying fungus on woods of various broad-leaved deciduous tree species all over the world. It is traditionally used as food after collecting from wild habitats for many centuries in Japan. Later, in 1928, its artificial cultivation technology has been developed with the use of sawdust and rice bran (Tonomura 1978, Nakamura 1981). Nowadays, it is widely cultivated in almost all East Asian countries. As regards the global production of this mushroom, it ranked 6th in 1997 with annual production of 130,000 mt, sharing 2.1 per cent of the total mushroom production (Chang 1999). The winter mushroom is particularly known for its taste and medicinal properties. It cures liver diseases and gastro enteric ulcers. Further, it contains immunomodulatory, antitumour, tumour inhibiting and antibiotic substances (Yoshioka et al. 1973, Ying et al. 1987, Sharma et al. 2009).

Morphology and Other Characters

The fruit body (sporophore) grows in clumps on dead wood, or on old stumps and decaying wood, erect or prostrate, small to medium sized, centrally stipitate, very viscid, velvet stemmed, several fruit bodies grow in cluster from a common and short root like structure. Pileus is 2–5 (6) cm in diameter, hemispherical or convex at first, later expanded to flattened surface at first hoary-pruinose, but soon polished and shining, viscid when moist, orange, yellowish brown, tawny or dark brown, margin light brown and inrolled, surface glabrous, viscid. Flesh is yellowish, white, soft, delicate, thin. Gills are white or light cream or yellowish white coloured, broad near the stipe, edge fringed under lens, adnexed or decurrent to the stipe. Stipe is central, stiff, stuffed or hollow, 2–9 (3–10) cm long and 0.2–0.8 cm (0.3–1.0 cm) in diameter, pale yellow when young, on maturation lower part dark brown, upper part light brown or white, lower half of stipe is covered with dense growth of reddish brown hairs giving velvety appearance, without volva and annulus. Basidia are 23–28 × 3.5–4.3 μm in size, four spored, hyaline in KOH. Basidiospores are white, narrowly elliptical or cylindrically oval with flat surface, non-amyloid, smooth, 5–7 × 3–4 μm (differently 7.6–9.0 × 4.0 μm) in size, spore in muss dull white. Pleurocystidia are 32–48 × 10–18 μm in size, scattered, fusoid-ventricose with obtuse apices, thin walled, hyaline. Cheilocystidia are similar to pleurocystidia, scattered. Hyphae are with clamp connexions (Smith 1978, Tonomura 1978, Purkayastha and Chandra 1985).

Facilities and Materials Required

Cold growing room condition: This mushroom is specifically suited for warm temperate areas and cold subtropical areas, since, it requires 10–12°C for fruit body formation. Sometimes, further low temperature (3–5°C) treatment is required for growing quality mushroom (Tonomura 1978). In subtropical plains of India this mushroom can be grown in winter if the above temperature prevails. However, for commercial cultivation, environmentally controlled growing room is required.

Polypropylene packet: Polypropylene packets of 2 kg capacity are required if cultivation is done in bag method.

Wide mouth bottle: Wide mouth bottles of 800 ml capacity are required if cultivation is done in bottles.

Sawdust: Sawdust of broad leave trees is used as substrate. This may be used alone of single tree species or as a mixture of several broad leave trees. The sawdusts of *Cryptomeria japonica* and *Chhamaecyparis obtusa* are suitable for mixing and cultivation (Tonomura 1978).

Rice bran: This is required for substrate preparation.

Autoclave: Large sized autoclave is required for sterilization of substrate mixture.

Thick paraffin paper or plastic film: This is required for giving support to fruit bodies.

Spawn

Both sawdust and grain spawns are used for cultivation of this mushroom.

Sawdust spawn

Formula 1: Sawdust and rice bran are mixed together in 10:1 ratio and added water to give 60–65 per cent moisture (Tonomura 1978). The mixture is then filled in spawn bottles or polypropylene bags for sterilization at 15 lbs p.s.i. for an hour. The inoculation is made by mycelia grown either in PDA medium or Malt Extract medium.

Formula 2: Sawdust of *Eucalyptus* sp. (80 per cent) and rice bran (20 per cent) with water 60 per cent of the mixture are used as spawn medium (Yang 1986). This mixture is filled in bottles and sterilized, and later, inoculated in the same method as done earlier. Incubation is done at 24ºC in dark for spawn growth. The spawn becomes ready for use after 20 days of incubation (Sharma et al. 2009).

Grain spawn

Boiled wheat grains are mixed with $CaCO_3$ (0.5–1.0 per cent) and $CaSO_4$ (1.5–3.0 per cent). The mixture is then filled in spawn bottles or polypropylene bags. These are plugged with non-absorbent cotton and sterilized in autoclave at 20 lbs p.s.i for 2 hrs. The inoculation with mycelial mat is done aseptically and incubated at 24ºC in dark for 20 days.

Cultivation

The cultivation of winter mushroom has been started in India in the year 1980 on oak sawdust and poplar leaves mixed with rice bran in small bottles or poly bags (Rangad and Jandaik 1980). Till date, its cultivation is mainly confined to different laboratories in our country.

Generally, this mushroom is cultivated on the medium made of by mixing sawdust of broad leave trees (80 per cent) and rice bran (20 per cent). The water content is adjusted in the medium to 58–60 per cent (Tonomura 1978, Kapoor 1989). In another way, sawdust is wetted with equal amount of water for 16–18 hrs. The wet sawdust is mixed with 10 per cent wheat bran (Sharma et al. 2006). The medium of either type is filled in polypropylene bags (2000 g capacity) or polypropylene bottles (540–650 g substrate in 800–1000 ml capacity) and sterilized in autoclave at 15–20 lbs p.s.i. for 1 hour. The sterilized substrate medium is now kept for one day for cooling. The cooled medium is inoculated by 2–3 per cent spawn aseptically and closed the mouth with cap or cotton plug. This is now incubated in dark in growing room at 22–25°C for spawn run. The mycelia cover the whole substrate within 20–25 days. When the mycelia spread to 90 per cent of the substrate surface, the mouth is opened and the spawn material which inoculated in the bag/bottle is removed by scrapping the upper portion of the substrate. The upper surface is made smooth by hand palm. The bags/bottles are kept in dark at 10–12°C where, humidity is maintained to 80–85 per cent. Primordial of fruit bodies develop within 10–14 days after reducing the temperature. At 10–12°C the fruit bodies grow faster, but they are slender, long, and of poor quality. Thus, to control the growth of mushroom the temperature is minimized further to 3–5°C, providing airflow @ 3–5 m/sec. This period, called as **controlling period**, lasts for about 5–7 days. During this period the humidity in the room is maintained to 70–80 per cent. When the fruit body reaches to the length of 2 cm, the neck of the bottles are wrapped with thick paraffin paper or with plastic sheet to give the support to the fruit bodies. While, in case of plastic bags, the folded neck is unfolded to give support. Next stage is growing or cropping period. At this stage temperature is again increased and that maintained at 5–8°C with room humidity 75–80 per cent. This period takes about 5–7 days for maturation of fruit bodies.

Harvesting and Packing

When the fruit bodies become 13–14 cm in length, these are pulled out by hand from the bottle or bag. It takes 50–60 days from spawn inoculation to fruiting. The fruit bodies are packed in polypropylene bags or can be sun dried. After harvesting, the same bag gives production at 15 days interval. About 180–240 g fresh mushroom is harvested in two flushes from 540 g substrate of a bottle. In case of polypropylene bag method, 360–400 g fresh mushroom is harvested from 2 kg substrate (Tonomura 1978, Sharma et al. 2009).

GIANT MUSHROOM CULTIVATION

The cultivation of giant mushroom, *Stropharia rugoso-annulata*, was initiated in West Germany (GDR) and then that was started in Poland, Czechoslovakia, Hungary and other European countries (Puschel 1969, Kindt 1971, Szudyga 1973, Stanek 1974, Balazs 1974, Zadrazil and Schliemann 1975). At present, it is one of the important commercially grown mushrooms in the world with an annual production of 1300 tons (considering the production year 1975, declaire 1978). The mushrooms are easy to cultivate with simple inexpensive methods. The mushrooms have certain advantages and disadvantages in cultivation. As regards the advantages, they are resistant to diseases and pests and can withstand in adverse climatic conditions (Szudyga 1978). On the contrary, the disadvantages, like unstable yield, fragile fruit body with hollow stipe and poor storage quality due to regrowth of mycelium from basidiocarp are very prominent in these mushrooms. In India their cultivation technology has been standardized by Upadhyay and Sohi (1989). The mushrooms are mildly scented with bland taste. The young fruit bodies are used for food preparation. They are suitable for preparation of stew, grill and pickle. The mushrooms contain 22 per cent protein on dry weight basis, where dry matter is 8 per cent. The niacin level is 3.41 mg.

Morphology and Other Characters

The sporophore is centrally stipitate. Pileus is 5–40 cm in diameter, colour variable on the variety and temperature of environment, margin regular with remnants of veil. In general, the colour of the young one is white which turns to yellow or brown with a reddish tinge in age. The young fruit body contains characteristic papillae. At low temperature the fruit body remains white in colour on maturation also. The lamellae are gray when young and bluish black when old. The stipe is round, with annulus and remnants of veil of cotton wool consistency at upper part; centre is hollow when mature, whitish cream in colour and thick at base. Volva is absent. The spore is bluish black in mass.

Cultivation

The mushroom can grow in temperature range 4.5° to 32°C. It requires 4–12 months after inoculation to form fruit body. Afterwards, it grows in flushes for 6 months to 2 years depending upon the substrate nature and types of maintenance. It is cultivated both in out door and in bag systems as follows:

Outdoor cultivation

Site selection: The mushroom beds are laid in partially shaded warm places in a protected area that sheltered from the wind.

Substrate preparation: The agricultural residues of cereals, like paddy, wheat, oat, flax, etc. are used as substrates of this mushroom. The fresh straw of any kind, either intact or chaffed, is placed in heaps and watered for 6–10 days by sprinkling water with intermittent turning for uniform soaking. Alternatively, the straw is dipped in water in a tank for about 48 hrs. The excess moisture is removed by drying in wind or sun to keep 70–75 per cent moisture during bed preparation.

Bed preparation: The bed is prepared with the moist straws placing in layers on a garden frame that prepared by digging out soil up to 30 cm deep in 1 square metre area. The straws are compressed by treading. The bed area as mentioned can accommodate 20–30 kg dry substrate.

Spawning and mushroom growing: The spawning is done by inserting bits of spawn in to 5–8 cm depth of the straw at several places keeping a little gap in between. Alternatively, spawn may be broadcasted over the straw surface before placing the last layer. After spawning, the bed is covered with straws of 7–8 cm thickness. The beds are kept as such for about 4–5 weeks for spawn run. After the said period, observation is needed to estimate the spawn run position. This can be done by seeing the mycelial ramification on the moist straw. In most cases, the upper layer of straws is dried up more to grow mycelia. At that time, the completion of spawn run is estimated by seeing below the upper layer of straws. After completion of spawn run, the beds are covered with forest soil (with humus) for casing. At that time, the upper dried up layer is removed and casing is done on the myceliated straws.

The cased beds are now kept as such for fruit body development. The initiation of fruit body starts after 7–8 weeks of spawning. In temperate zone, where the mushrooms are mostly cultivated, the fruit body develops in the month of August, and their production continues up to the onset of frosty weather. The mushrooms require 10–12 days for their development from pinning.

Bag cultivation

The bag cultivation is done in polypropylene bags or polyethylene bags as practiced for milky mushrooms with certain alteration.

Spawn: Spawn is used, either wheat grain spawn or straw spawn.

Substrate preparation: Different agricultural residues, such as cereal straws, flax straws, corn cobs, sugarcane bagasse are chopped into small pieces. They are water soaked and removed the excess water by decanting and drying. The substrates are sterilized in autoclave or pasteurized by steam or disinfected by hot water treatment. The excess water and moisture, if any, are removed as followed earlier. The moisture in substrate is kept 55–60 per cent during bag filling (Upadhyay and Sohi 1989).

Spawning and cultivation: The disinfected wet straw is mixed with 1 per cent wheat grain spawn and filled in polythene bags. The spawn may be applied in layers by preparing substrate layers inside the bag. When the bag is filled up with the spawn and substrate, the mouth of the bag is closed and kept inside a house under dark condition at 22–26°C for 20–30 days. This incubation is done for spawn run in the substrate. On completion of spawn run, the mouth of the bag is opened and the myceliated substrate is cased with sterilized or pasteurized forest litter and FYM mixture (1:1 ratio) up to a height of 2.5 cm. The cased bags are now kept in a growing room at 80–90 per cent relative humid condition. Indirect sunlight for 10–12 hrs/day is allowed to enter in the room. Watering is done as and when required. The fruit body comes out within 8–50 days after casing and that matures within 10–12 days.

Harvesting

The mushrooms are harvested by hand plucking, giving a gentle twist at base region of the fruit body and simultaneous pulling. About 12 kg of mushrooms are produced per 1 square metre of bed area in out door cultivation. In case of bag cultivation similar method is applied to harvest mushrooms, but here the gap which appeared after harvesting is filled up with casing soil. About 275–750 g of mushroom is produced per bag of 2.5 kg straw substrate.

PARASOL MUSHROOM CULTIVATION

The parasol mushroom *Macrolepiota procera,* grows solitary or in clusters in soil, pastures, lawns, gardens and woods. Generally, this mushroom is marketed in small scale collecting from local habitats. In this chapter, a trial cultivation is presented to emphasize the possibility of its commercialization.

Morphology

The sporophore is about 12–20 cm long. Its pileus is 7.6–15.2 cm in diameter, at first ovate, later campanulate and finally becoming expanded with a reddish brown umbo. The colour of the pileus is grayish brown to reddish brown; surface is cracking into large brown scales that arranged concentrically. Gils are free, crowded, white when young and pinkish on maturity. Stipe is slender, cylindrical, with small scales, hollow, with a thick and tough annulus. The base of stipe is bulbous and without volva. Flesh is white, remaining unchanged on cutting, soft and thick. Basidium is four spored and clavate. Basidiospores are white, ovate to elliptical, apiculate, smooth, with germ pore and 13.0–16.5 × 8.0–10.0 μm in size. Spore in mass is white. Pleurocystidia is absent (Purkayastha and Chandra 1985, Biswas 1988).

Facilities and Materials Required

The facilities and materials as required for milky mushroom cultivation are required. These are as follows:

1. Mushroom house or shed
2. Wooden or bamboo racks
3. Hand chopper/chaff cutter—1
4. Metallic drum for water boiling
5. Plastic bucket—2
6. Bamboo cane basket—2
7. Sprayer
8. Gunny cloth
9. Fire wood or large gas oven
10. Polypropylene bags (18″ × 14″)
11. Polythene sheet (for spreading wet substrate in removing excess moisture)
12. Nylon thread
13. Nylon net
14. Substrates: Fresh one year old paddy straw of *aman* paddy and pea haulms spawn
15. Fungicide
16. Antibiotics
17. Insecticide
18. Garden soil
19. Farm Yard Manure (FYM)
20. Sand

Cultivation

Processing of substrate

Substrates, like, paddy straw or pea haulms are cut into 2–4 cm long pieces with a hand chopper or chaff cutter. These are steeped in water and disinfected by hot water treatment. In this method, the substrate is water soaked for 16–18 hrs for overnight in a large container. On next day morning, the substrate is kept on a raised platform for about 1 hour to remove the excess water. After that, the water soaked substrate is kept in a large bucket and there boiled water is poured so that the substrate is fully immerged in hot water. The whole set is covered by a wooden plank and kept for 2 hrs. Time to time boil water is added to maintain the temperature of above 65ºC. The substrate is then taken out in a bamboo cane basket for draining out the excess water and later spread in sun to remove excess moisture for about 1–2 hrs depending upon the brightness of sun light. By this way, moisture is kept

60–70 per cent in the substrate and that can be visually estimated by seeing the surface layer of substrate which appeared as dry

Preparation of mushroom bed

The process of parasol mushroom cultivation is same as milky mushroom cultivation in polythene bag. In this method, transparent polythene bag (18″ × 14″) is perforated (hole size 0.4″ diameter) at a distance of 4″ all over its body surface. Disinfected paddy straw (about 0.75 kg on dry weight basis) as made ready after disinfection and a half bottle/bag of spawn (75 g) are used to prepare a bag. The substrate is placed inside the bag in 4–5 layers by pressing with hand palm. After completion of one layer 10–15 g of spawn grains are spread over the substrate layer. Then, another layer of substrate is prepared and the spawn grains are sprayed over that similarly. In this way 4/5th portion of the transparent polythene bag is filled up by the spawned substrate. Now, the bag is closed by tying with a piece of nylon thread and incubated in spawn running room in dark for about 18 days.

Casing material

During spawn running stage, casing soil is made ready by mixing field soil, sand and well decomposed FYM in 1:1:1 ratio. The mixture is then suspended in boiling water for an hour. Then, the excess water is removed by decanting the whole material on a clean surface. The material is kept as such for drying enough in order to make it fragile by hand palm pressure

Casing

When the spawn running is completed in polythene bag, the mouth of the bag is unthread or cut open and the casing soil as prepared earlier is spread over the spawn run substrate up to 1 inch of thickness.

Fruit body production

The mushroom bed is now kept on a rack for allowing fruit body production repeatedly from the upper surface (Figure 7.7B and C). Generally, it takes 60–75 days for harvesting of fruit body from the date of inoculation of spawn into the substrate in polythene bag.

Harvesting

Harvesting is done when the pileus remained as round, i.e. when the edge of pileus still attached with annulus ring by a veil (a thin membrane like structure).

Chapter 8

Management of Diseases, Pests and Weed Fungal Attacks

Like crop plant, mushrooms are also affected by different diseases, pests and weeds (moulds) during cultivation (Figures 8.1–8.4). The occurrence of the diseases in mushrooms is the results of both primary and secondary infections. A minor infection in compost, casing soil or other substrates as used for mushroom cultivation may cause severe outbreaks of diseases in mushrooms. The spawn is considered as the main source of viral diseases. The vegetation surrounding the mushroom farms harbours mushroom pests. The wind borne dust particles and contaminated tools, machinery and clothing are also the sources of several pathogens and weed fungi. The occurrence of diseases and pests of mushrooms is mostly prevented by heat treatment. Since, most of them can easily be killed by heat treatment at 65°C for 30 minutes (Hayes 1978). Thus, heat treating of the compost, casing soil and other substrates is the general practice in mushroom cultivation. The tools and machineries are generally disinfected with formalin to avoid contamination. In spite of that, several times the diseases, pests and weed fungi are found to cause considerable damage of mushroom, which may be due to improper cultivation practices, and thus proper care is needed to prevent their attacks.

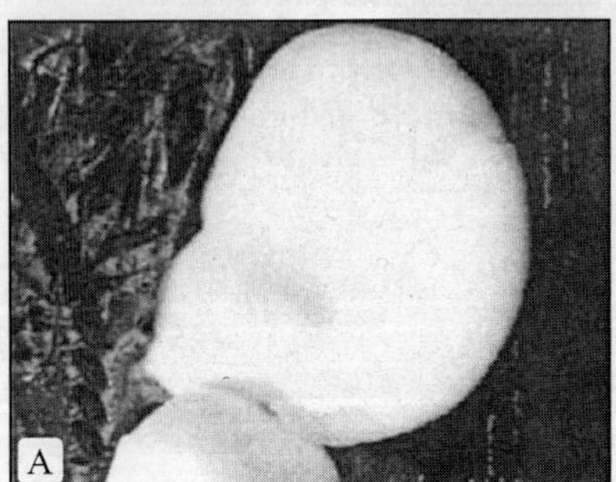

Figure 8.1 (*Contd.*)

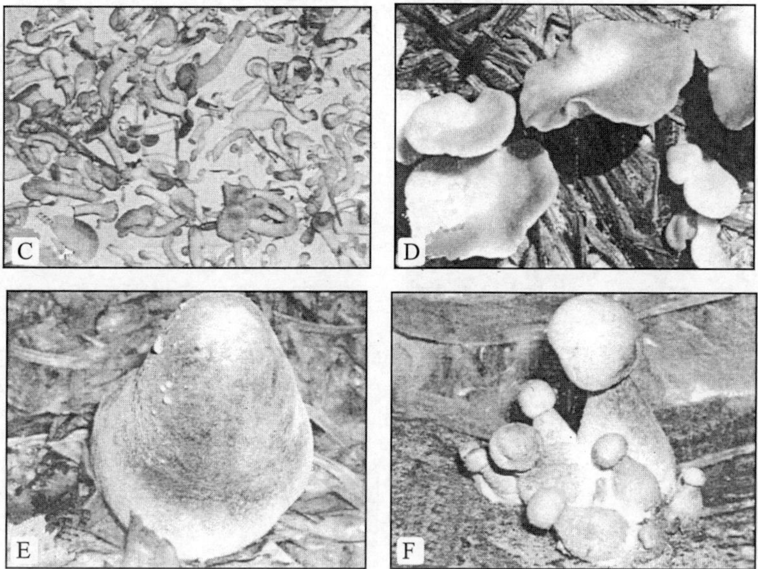

Figure 8.1 Diseases of mushrooms. **A.** Yellow blotch in oyster mushroom, **B.** Yellow rot in oyster mushroom, **C.** Bacterial rot in oyster mushroom, **D.** Brown blotch in oyster mushroom, **E.** Bacterial rot in paddy straw mushroom and **F.** Brown blotch in milky mushroom.

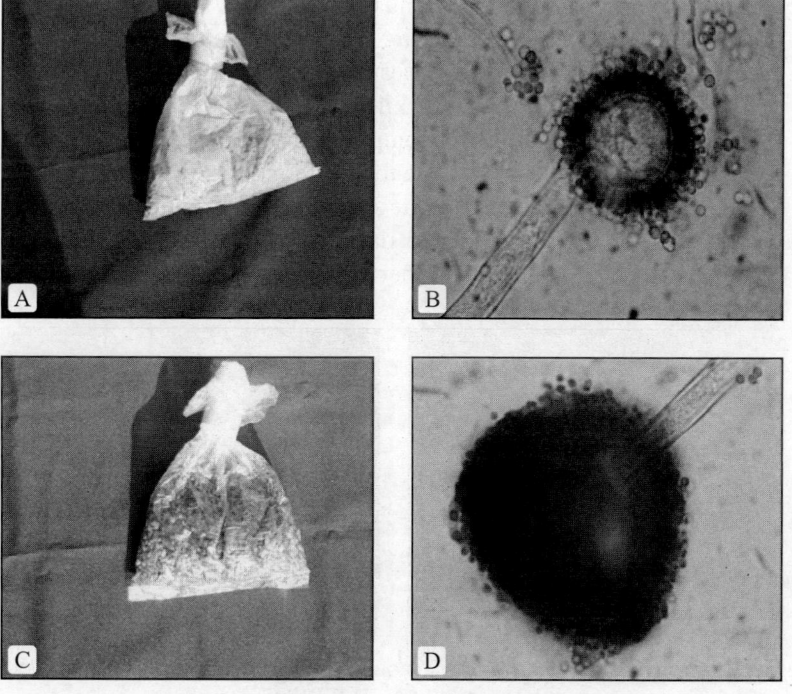

Figure 8.2 (*Contd.*)

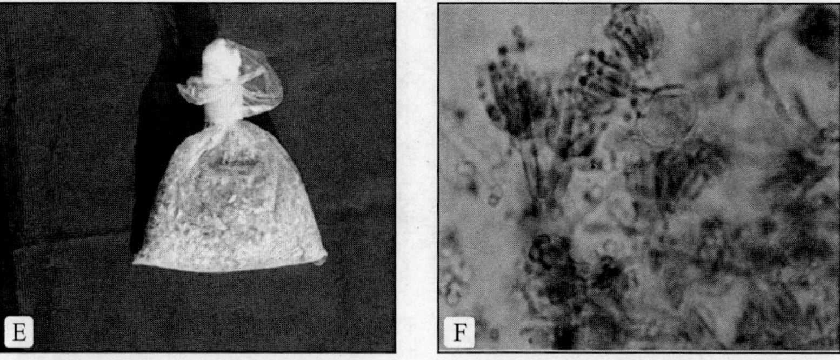

Figure 8.2 Mould in spawn medium. **A.** Yellow mould, **B.** *Aspergillus flavus* causing yellow mould, **C.** Black mould, **D.** *Aspergillus niger* causing black mould, **E.** Blue green mould and **F.** *Penicillium* sp. causing blue-green mould.

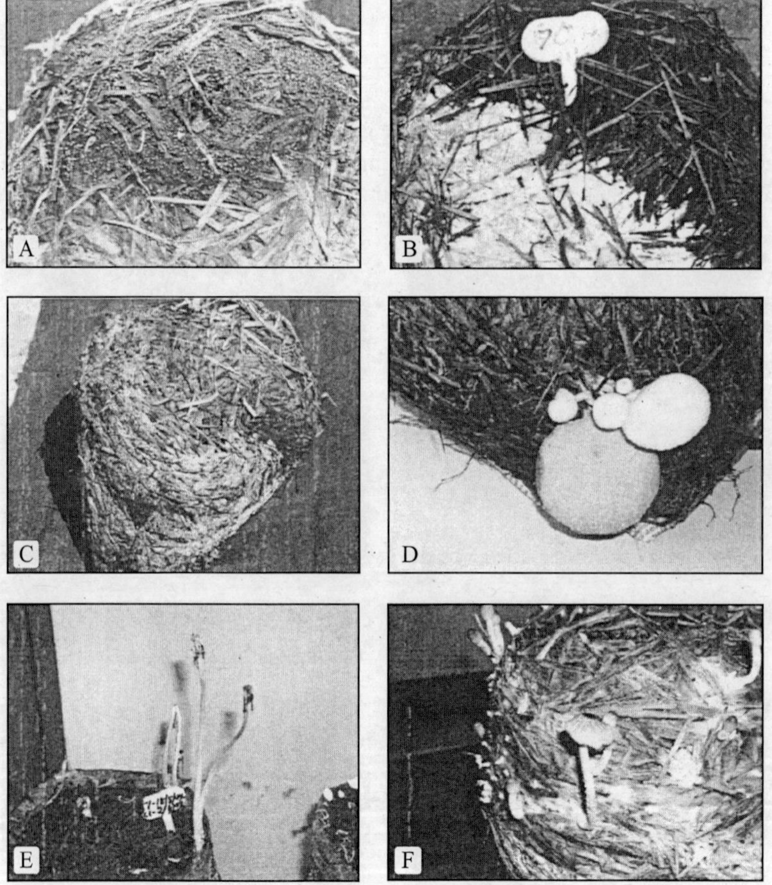

Figure 8.3 Moulds in mushroom bed. **A.** Dark mould, **B.** White mould, **C.** Yellow mould, **D.** Yellow mould covered fruit body, **E.** Inky caps and **F.** *Coprinus* sp.

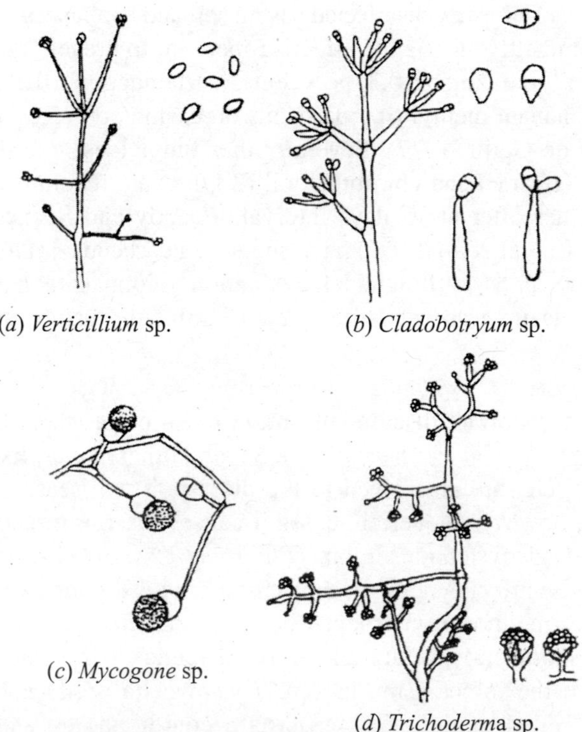

(*a*) *Verticillium* sp. (*b*) *Cladobotryum* sp.

(*c*) *Mycogone* sp. (*d*) *Trichoderma* sp.

Figure 8.4 Major disease causing organisms.

DISEASES OF BUTTON MUSHROOM

Fungal Diseases

Dry bubble

The dry bubble disease is very common in button mushroom (*Agaricus* sp.) cultivation. It is also known as *Verticillium* disease, brown spot, fungal spot, La mole, etc. The disease is mostly caused by *Verticillium fungicola.* Several other species of *Verticillium*, like, *V. malthousei, V. psalliotae, etc.,* are also found to cause the disease. The disease appears in mushroom at first in forming onion shaped fruit body with stipe broader than the cap which later become crooked and deformed with distorted stipe and tilted cap. The colour of mushroom changes to grey with localized light brown depression. Finally, the diseased fruit body shrinks in blotched areas, turns leathery, dry and shows cracking. Further, the fruit body is covered with fine grey white mycelia and forms pimple-like outgrowths on its surface. When a part of the pileus is affected, hare-lip symptom is developed. The main source of infection is the contaminated casing soil. The contaminated casing soil shows whitish patches, which later turns to grayish, due to mycelial growth of the causal fungi. In certain cases, contaminated appliances and hands are also found to cause infection. The phorid and sciarid flies are mostly responsible for secondary infection in mushrooms.

Management: The use of properly disinfected casing soil and appliances as well as maintenance of hygiene in mushroom cultivation is the effective measure to prevent the disease occurrence. Spraying of fungicide, like zineb (0.3 per cent), carbendazim (0.1 per cent), benomyl (0.15 per cent) or thiophanate methyl (0.2 per cent) on casing soil also minimizes the disease (Seth et al. 1973, Van de Geijin 1977). Several, other fungicides, like triadimefon (1 g/m^2), prochloraz manganese (1 g/m^2) and chlorothalonil (2 g/m^2) are recommended to apply within 7 days of casing and thereafter at 15 days intervals (Gandy and Spencer 1981 Van Zaayen and Andrichem 1982). Katyal et al. (2006) have suggested carbendazim to manage the disease, since, in a comparative *in vitro* fungicidal evaluation thiophanate methyl was not much effective while; carbendazim was very effective to inhibit the pathogen.

Wet bubble

This is one of the most important diseases of button mushroom, which caused by *Mycogone perniciosa* (perfect stage: *Hypomyces perniciosa*). Several other names, like *Mycogone* disease, white mould, La Mole, etc. are also given to the disease. The disease mostly affects button mushroom cultivation, however, in certain cases it causes disease in paddy straw mushroom (*Volvariella volvacea*) and oyster mushroom (*Pleurotus eryngii*) (Chang 1978, Sharma and Kumar 2000). The disease affects young mushrooms at pin head stage and due to malformation of stipe and pileus it forms monstrous shaped body which often does not show resemblance with mushroom. The short, curly, white and fluffy mycelia of the causal fungus can easily be seen on and around the infected mushroom. The mycelia produce two kinds of spores, such as hyaline one celled conidia on verticillate conidiophores and two celled brown chlamydospores. At latter stage, the fruit body becomes completely rotten and exudes brown liquid drops on its body with emition of foul odour (Hayes 1978, Sharma 1995). The chlamydospores survive in soil for years. Thus, the soil with chlamydospores is the major source of infection (Hayes 1978). The dissemination of the disease occurs through air, water splash, mites and flies (Hsu and Han 1981, Garcha 1984). The pathogen is very much sensitive to heat with thermal death point 48ºC (Van Zaayen and Rutigens 1981).

Management: The use of properly pasteurized (at or above 65ºC for 1 hr or more) casing soil, disinfected appliances and maintenance of hygienic condition in mushroom cultivation are the effective measures to prevent the disease occurrence. Further, spraying of fungicide, like, zineb (0.3 per cent), carbendazim (0.1 per cent), benomyl (0.15 per cent), thiophanate methyl (0.2 per cent), chlorothalonil (0.2 per cent) or prochloraz manganese (0.1 per cent), immediately after casing or mixing them in proportionate amounts with casing soil also controls the disease very well (Stanek and Vojtechovska 1972, Kim 1975, Hsu and Han 1981, Fletcher et al. 1983, Sharma 1995, Sharma and Kumar 2000, Singh and Sharma 2001).

Cobweb disease

The disease is caused by *Cladobotryum dendroides* (Syn: *Dactylium dendroides*; perfect stage: *Hypomyces rosellus*). The disease is also known as **mildew**, soft decay, *Hypomyces* mildew disease, *Dactylium* disease, etc. It appears mostly in high humid condition. The appearance of the disease is first noticed as small circular white patches of mycelial growth on casing soil or substrate. The floccose white mycelia later cover the entire body of mushroom.

The affected mushroom then becomes died and rotten, showing brown in colour which finally turns to pink or red. In severe case, the causal fungus develops as fluffy cobweb like dense mat of mycelium on casing layer and mushrooms. The pathogen is a soil borne fungus and the infection occurs by the contaminated soil or by conidia which can easily be spread by air.

Management: The use of properly pasteurized (at or above 65°C for 1 hr or more) casing soil, disinfected appliances and maintenance of hygiene in mushroom cultivation are effective measures to prevent the disease. Application of zineb (0.3 per cent), carbendazim (0.05–0.1 per cent), benomyl (0.15 per cent), chlorothalonil (0.2 per cent), prochloraz manganese (0.1 per cent) immediately after casing or mixing them in proportionate amounts with casing soil also gives very good control of the disease. Dusting of zineb or mancozeb on mushroom bed @ 100 g per 100 m^2 in between two flushes and spraying of weak solution of formalin (0.2–0.3 per cent) are effective preventive measures of the disease (Sharma 1995). Prevention of disease can also be done by applying bavistin (carbendazim) and TMTD (tetramethylthiuram disulfide/thiram) mixture at 0.9 and 0.6 g/m^2, respectively, on bed (Seth and Dar 1989).

Trichoderma spot (Green mould)

This disease, which is also called as *Trichoderma* blotch and *Trichoderma* mildew, is caused by several green mould species of *Trichoderma*, namely, *T. aggressivum*, *T. atroviride, T. haematum, T. harzianum, T. koningii, T. logibrachiatum T. pseudokoningii* and *T. viride*. However, amongst them, *T. harzianum* is most conspicuous in causing disease (Morris et al. 1995). In many cases, the disease is simply identified as 'green mould' based on the growth pattern of the causal fungi on substrates. The appearance of the disease is first evident by forming green mould of the causal fungus on substrate or casing layer. Later, it attacks the fruit bodies producing dark brown or greenish lesions on the pileus and stipe. In certain cases, the infected mushroom is cracked and distorted. If the infection is caused by *T. viride* then there a reddish brown discolouration is visible along the stipe. On the contrary, the pathogen, *T. koningii* grows as cottony weft of grayish mycelium over the substrate and whole mushroom (Sharma 1995). However, in all cases at advanced stage, the infected mushrooms rot as wet.

Management: Maintenance of proper hygiene at different stages of cultivation reduces the disease. Proper pasteurization and conditioning of compost are required to prevent the disease. Use of formalin (2 per cent) as spray on casing soil just after application of the latter on bed is effective to minimize the disease. Weekly spray of mancozeb (0.2 per cent) or carbendazim (0.1 per cent) or dusting of zineb on mould affected area is effective measure for the management of the disease (Bharadwaj and Seth 1983, Sohi 1988). Katyal et al. (2006) have suggested carbendazim to manage the disease, since, in a comparative *in vitro* fungicidal evaluation that was effective to inhibit the pathogen at very low (3–4 ppm) concentration.

Weed Fungi

Apart from the disease causing fungi, quite a large number of saprophytic fungi appear in mushroom beds as a result of improper or incomplete preparation of compost and casing soils. These moulds cause considerable damage to the crop in terms of reducing yield. Under certain conditions, these undesirable fungi compete for food in the beds with mushroom mycelia and crowed out the growth of the planted spawn. Such undesirable fungi are called **weed moulds**. These are as follows:

False truffle

False truffle, also known as truffle disease, is caused by *Diehiliomyces microsporus*. This is one of the most dreaded competitors in button mushroom beds (Lambert 1930). The appearance of the weed fungus is the major problem for cultivation *of Agaricus bitorquis* in India, since, the higher temperature (beyond 22ºC) which favoured the growth of mushroom also favours the growth of false truffle. The fungus also affects the cultivation of *A. bisporus* when the temperature rises up beyond 20ºC (Sohi et al. 1965). It does not allow mushroom mycelia to grow inside the compost. The weed fungus first appears as small wefts of white to cream coloured mycelia in compost and casing material. Later, it turns to solid wrinkled or round mass of fungal body (ascocarp) of size ranging from 0.5 to 3 cm in diameter. At first the solid mass is white later on maturity that turns to pink or reddish. The fungal body becomes dry and disintegrates into powdery mass emitting chlorine like odour (Sharma 1995).

Management: This weed fungus is very difficult to manage if proper preventive measures are not taken well in advance. Maintenance of proper hygiene in mushroom cultivation is needed in every step. The major sources of truffle are the casing soil, where it survived as ascospores, and wooden trays, where it survived as mycelium from the infected preceding crops. Thus, well disinfected or pasteurized casing soil is used in poly bag or self system of cultivation, discarding tray cultivation. In order to prevent contamination of the weed fungus from soil, the compost is prepared on a concrete floor. The compost is allowed to be well pasteurized and conditioned. In addition, in case of common button mushroom (*A. bisporus*), maintenance of controlled temperature is in practice. The temperature during cropping period is maintained below 18ºC to avoid the growth of the truffle. This manipulation in temperature has no use in *A. bitorquis* cultivation, since the above mentioned temperature is suited for the growth of this mushroom. However, in the latter case, the resistant strains like Horst K 26 and Horst K 32 are used to combat the truffle disease (Van Zaayen 1982). As regards, the chemical control of the truffle, none of the fungicides, like, benomyl, captafol, chlorothalonil, copper sulphate and mancozeb are effective to control the fungus in mushroom bed, although, benomyl inhibits the truffle *in vitro* study using malt agar as a medium. Further, the spores of the truffle are very resistant to formaldehyde, copper sulphate, mercuric chloride, sulphur dioxide, chlorine, lime, sulphur, chlorinated phenols and ceresol (Gandy et al. 1953, Van Zaayen and Vander Pol-Luiten 1979, Tewari 1987).

Olive green mould

Two different species of *Chaetomium* (*C. olivaceum and C. globosum*) are found to cause olive green mould in mushroom bed. The moulds appear after 10–12 days of spawning as grayish white mycelia on compost surface. The mycelia later form superficial fruiting bodies appearing as green cockle burns on straw in isolated patches of the affected compost. The affected compost shows musty odour and hinder the growth of mushroom mycelia. The fungus, *C. olivaceum* forms superficial perithecia that appeared as opaque, globose, thin, membrane with apical tuft of dark bristles or setae, while, *C. globosum* forms scattered and gregarious perithecia appearing as opaque, broadly ovate or ellipsoid, often pointed at base, thickly and evenly clothed with slender flexuous hairs (Sharma 1995).

Management: Occurrence of the weed fungi in mushroom bed can be well prevented with the use of quality compost. Thus, much care is needed in preparation of compost and its pasteurization. The nitrogenous ingredients, like chicken manure, urea, ammonium sulphate, etc. which is generally used for compost preparation, should not be added at later stage of composting. Sufficient time should also be given for proper fermentation of the compost ingredients. The pasteurization time should not be curtailed. Apart from this, if the mould appeared in mushroom bed, zineb (0.2 per cent) can be sprayed to avoid secondary contamination (Sohi 1986). Spraying of systemic fungicides, namely, benomyl, carbendazim or thiophanate methyl at 0.1 per cent is also effective to minimize the mould in bed (Thapa et al. 1979).

Brown plaster mould

Brown plaster mould, caused by *Papulospora byssina*, is reported to cause severe damage in both button and oyster mushrooms in northern parts of India (Munjal and Seth 1974, Dar and Seth 1981). The fungus appears on mushroom beds during cropping as extensive patches of colonies, inhibiting the growth of mushroom mycelium. At first, its appearance is noticed as whitish mycelia on the exposed surface of compost and casing soil as well as on the sides of bags. Later, it turns to large patches of light brown to cinnamon brown or tan coloured zones which finally become rusty coloured. The brownish mycelia of the fungus produce clusters of brown coloured spherical many celled sclerotia or bulbils.

Management: Management of the weed fungus is same as practiced for olive green mould. The compost is required to be well fermented and pasteurized. Strict hygienic conditions are to be maintained in every step of cultivation process. In case of button mushroom the substrate should be well pasteurized. Several fungicides, namely, benomyl, carbendazim and thiophanate methyl at 0.1 per cent are effective as spray on mushroom beds to minimize the mould (Sharma 1995).

White plaster mould

The white plaster mould fungus, *Scopulariopsis fimicola,* causes severe damage in white button mushroom yield at different places in India, particularly, when the cultivation is done in long method of composting (Kaul et al. 1978, Sohi 1986, Garcha et al. 1987, Bharaddwaj et al. 1989). The fungus initially appears as small irregular patches of white filamentous aerial growth which later becomes oppressed to the compost or casing soil and forms white patches of mycelial mat. The colour of mycelial mats changes from white to light pink with age. This mould forms chains of ovate to globose annellospores on short condiophores in clusters and significantly reduces the growth of mushroom mycelia in the compost. This weed mould is usually associated with wet and greasy conditions of the compost which retained traces of ammonia odour.

Management: As the fungus occurs on poor quality compost with high ammonia and pH, thus, care is needed in preparing good quality compost with proper decomposition and pasteurization. Addition of gypsum during the composting is essential to control white plaster mould. The white plaster mould is usually associated with the dispersion of colloidal material in the compost which derived from the excess of ionized potassium. Gypsum (calcium sulfate) in compost serves to flocculate these colloids. Care is also needed to remove ammonia by spreading the compost in air for one or more days if smelling, indicated ammonia's presence, and to maintain proper pH giving correct quantity of compost ingredients as per a given formula and water. Spraying of benomyl (0.1 per cent) and local application of formalin (4 per cent) after removal of mycelial mat of the fungus are also helpful in managing the mould in mushroom beds.

Yellow mould

Several fungi, like *Chrysosporium luetum, C. sulphureum, Myceliophthora lutea, Sepedonium chrysosporium* and *S. maheshwarianum* are found to cause yellow mould in mushroom beds. The yellow mould caused by *Sepedonium* sp. is generally referred as '*Sepedonium* yellow mould', while the moulds caused by others are referred as yellow mould, mat disease or Vert-de-gris. The yellow mould fungi form yellow brown corky mycelial layer in the compost just below the casing (mat disease) or anywhere in the compost as circular colonies (confetti) or throughout the compost (Vert-de-gris). The *Sepedonium* yellow mould develops in the lower layers of the compost initially as white mycelial specks, which later changes to yellow or tan coloured at maturity.

Management: Proper pasteurization of casing soil and compost is the best measure to manage the yellow moulds in mushroom beds. In addition, maintenance of hygienic conditions during every step of cultivation processes is essential. The appliances and racks are to be disinfected with formalin (2 per cent). Spraying of carbendazim or benomyl (0.04–0.05 per cent) or calcium hypochloride (15 per cent) is effective to control yellow mould in mushroom beds. Further, incorporation of carbendazim (0.1 per cent) in compost is effective to prevent yellow mould in mushroom beds (Sohi 1986, Seth and Bhardwaj 1989, Vijay et al. 1993).

Cinnamon brown mould

The mould is caused by *Chromelosporium fulva* (perfect stage *Peziza ostrachoderma*). It appears on compost or casing layer initially as large circular patches of white mycelia. Later the colour changes to yellow gold to golden brown or cinnamon brown. The fungus produces numerous small fleshy discoid and golden fruit bodies.

Management: Use of properly pasteurized compost and casing soil is the best measure for the management of the mould. Care should also be taken to avoid excess moisture in the bed. If appearance of the mould in mushroom bed has already been established, then localized spraying of zineb (0.2 per cent) after removal of the affected compost has some beneficial effect.

Pink mould

The fungus *Cephalothecium roseum* causes pink mould in mushroom bed (Seth 1977, Kaul et al. 1978). It appears on casing soil as white mycelial growth, later, the colour changes to pink with age. The mycelia of the fungus produce erect conidiophores with acrogenous pear shaped, two celled and hyaline to pink conidia. The occurrence of the mould in mushroom bed generally happens through air.

Management: Spraying of thiram or captan (0.04 per cent) on casing soil twice at 10 days interval is effective to manage the disease (Guleria and Seth 1977).

Black whisker mould

This fungus, *Doratomyces* sp., produces black powdery spores, which appeared as smoke when disturbed. The mould appears in compost if that overheated during spawn run. Simple carbohydrates are mostly utilized by this fungus, but it can also utilize lignin.

Management: Use of properly pasteurized substrate and casing soil, and avoidance of excessive humidity, high temperatures and poor ventilation in growing room are effective measures for the mould. Spraying of carbendazim (0.05 per cent) or zineb (0.02 per cent) is also effective to control the mould when appeared in mushroom bed.

Blue-green mould

Blue green mould is caused by several species of *Penicillium*. Abundant blue-green spores are produced on the surface of the substrate. The mould appears in compost when spawn run occurred in high temperature. It utilizes simple carbohydrates, cellulose, starch, fat, and lignin. It is very common in cultivation of *Agaricus bitorquis,* where it is one of the chief concerns. Spores are airborne and ubiquitous.

Management: Use of properly pasteurized substrate and casing soil, and avoidance of excessive humidity, high temperatures and poor ventilation in growing room are effective measures for the mould. Spraying of carbendazim (0.05 per cent) or zineb (0.02 per cent) is effective to control the mould in mushroom bed.

Black mould (also yellow mould and others)

Several species of *Aspergillus* of various colours are grown on compost and in culture and spawn medium. Well used wooden trays and shelves for holding compost are frequent habitats of this contaminant in the growing house. The compost prepared in long method is a subject to harbour the mould. Different species of *Aspergillus* are found to cause various coloured moulds in mushroom beds. For example, *Aspergillus niger* produces black mould, while *A. flavus* produces yellow mould, *A. clavatus* produces blue-green mould, *A. fumigatus* produces grayish green mould and *A. veriscolor* exhibits a variety of colours (greenish to pinkish or yellowish). These moulds change in colour and appearance according to the medium on which they occur. Several species are thermophilic.

Management: It is necessary to sterilize culture and spawn media at 15 lbs p.s.i. and 20 lbs p.s.i., respectively, to avoid contamination in culture and spawn. Use of properly pasteurized substrate and casing soil, and avoidance of excessive humidity, high temperatures and poor ventilation in growing room are effective measures for the mould. Spraying of carbendazim (0.05 per cent) or zineb (0.02 per cent) is also effective to control the mould in mushroom beds.

Pin moulds

The mould is caused by *Rhizopus* sp. It is a very fast growing fungus. Once it sporulates, it forms many tall aerial hyphae adorned with black-headed pins. It grows on readily available carbohydrates. The mould is the primary contaminants of grain spawn. It is also very common on unpasteurized compost made up of straw.

Management: Use of well sterilized culture and spawn media is necessary to avoid contamination in culture and spawn. Well pasteurized substrate, casing soil, and avoidance of excessive humidity, high temperatures and poor ventilation in growing room are effective to control the mould. Spraying of carbendazim (0.05 per cent) or zineb (0.02 per cent) controls the mould in mushroom bed.

Inky caps

Different species of *Coprinus*, viz. *C. atramentanium*, *C. commatus*, *C. fimetarius* and *C. lagopus*, are found to appear in mushroom beds due to non-pasteurization or improper pasteurization of the compost. These weed fungi are also very common in paddy straw mushroom, milky mushroom and oyster mushroom cultivation where fresh agricultural residues, particularly paddy straw, are used as substrates. In most cases, more nitrogenous supplementations and high humid conditions augment their appearance. Inky caps appear as cluster of slender bell shaped sporophore in mushroom beds during spawn run and in newly cased beds during mycelial ramification. The sporophore of the inky cap is at first cream coloured, later changes to bluish black and usually covered with white scales which disappear within a few hours. On maturity, the inky caps decay due to auto digestions forming a blackish slimy mass.

Management: Use of properly pasteurized compost and casing soil is effective to minimize the appearances of inky caps in mushroom beds. Excess watering should be avoided. Young fruit bodies of *Coprinus* sp. may be removed by hand picking.

Lipstick mould

The mould is caused by *Sporendonema purpurescens* and frequently appeared in Punjab and Himachal Pradesh (Sohi 1986, 1988, Garcha et al. 1987). At first, the mould appears in mushroom bed as white crystalline specks, which later changes to pink, cherry red, dull orange and buff coloured mycelial mat, subsequently on maturity of spores. Soil, spent compost and casing mixture are the sources of primary inoculum. Spread of the mould occurs by water splashes.

Management: Use of properly pasteurized casing mixture and compost is the best measure to prevent the mould attacks in mushroom bed. Further, adoption of proper sanitation measures for cleaning mushroom house and appliances before cultivation as well as maintenance of hygiene during cultivation is essential.

Liliputia mould

The competitor mould *Liliputia rufula* (Perfect stage: *Gliocladium prolificum*), seriously restricts spawn growth in compost. It appears in mushroom bed during spawn run stage and affects the yield considerably both in hills and plains of northern India (Seth and Munjal 1981). Generally, the compost prepared with the use of horse manure or chicken manure is affected by the mould.

Management: Application of fungicide, like zineb at 0.002 per cent is effective to control the mould without hampering the growth of *Agaricus bisporus*.

Cat ear fungus

The fungus *Clitopilus passeckerianus* has been found to appear in mushroom beds in Kashmir valley as a weed fungus, forming white cat ear shaped fruiting bodies in tufts (Kaul et al. 1978, Garcha et al. 1987).

Management: Use of properly pasteurized compost and casing material prevents the mould. If the mould appeared in a mushroom bed, then hand picking is the only measure to control the mould.

Peziza

The competitor mould *Peziza* sp. is found to occur frequently in mushroom beds as buff coloured or dull-yellow coloured small thaloid fruiting structures in clusters (Kaul et al. 1978, Garcha et al. 1987).

Management: Excess watering is to be avoided. Pasteurization of compost and casing soil is effective to prevent appearance of the mould in mushroom beds.

Bacterial Diseases

Brown blotch (Bacterial blotch)

The brown blotch disease is caused by *Pseudomonas tolaasii* (= *P. fluorescens* biotype G) and is the only bacterial disease of button mushrooms reported from India (Fletcher et al. 1986, Sharma 1995). In India it causes 5–7 per cent loss of white button mushroom yield, although, in severe cases 60–70 per cent loss may occur (Guleria 1976, Thapa and Seth 1987). The main source of the disease in mushroom beds is the casing soil. The disease is favoured by dampness due to high humidity (RH over 85 per cent), high temperatures (over 20ºC) and poor ventilation in the growing room. It can also be spread by the workers' hands, irrigating water and on the inventory. Mushroom flies and mites are also responsible for the spread of the disease. The disease appears on the moist areas of caps and the areas where mushroom caps are touched with contaminated hands or tools. The pathogen induces lesions in mushroom tissue initially as pale yellow which later changes to golden yellow and finally to chocolate brown and slimy. This discolouration is superficial, not appeared more than 2–3 mm deep and the underlying mushroom tissue appeared as water-soaked, grey or yellow grey in colour. The disease can also cause distortion and splitting of the stipes. When the infection is severe, the spots spread throughout the whole mushroom surface. Seriously diseased mushrooms become deformed and the caps are decayed, giving a foul (unpleasant) odour. Young pins affected by the disease become brown and do not develop further. It is evident that this pathogen secretes a toxin which results in superficial lyses of cap tissue.

Management: It is necessary to maintain strict hygiene and sanitation in mushroom cultivation. Use of properly prepared substrate and casing soil that have been adequately pasteurized is needed. Avoidance of excessive humidity, high temperature and poor ventilation in growing room is essential. Precaution is to be taken to avoid fluctuation of temperatures in the growing room, since, that may cause water condensation on mushroom caps. Removal of diseased mushrooms which are the sources of the disease is to be done. Care should also be taken to control mushroom flies and mites which can spread the bacteria. Spraying of chlorinated water (125 ml of 10 per cent chlorine solution in 100 litre of water per 100 m^2 of bed area), streptomycin (0.02 per cent), oxytetracycline (0.03 per cent) or agricultural antibiotics, like pushamycin, plantamycin, agrimycin, etc. at 0.05 per cent of commercial formulations or application of terramycin at 9 mg/sq. ft. before first break, i.e. when the pin size is 4–5 mm, are effective to control the disease. Further, precaution should be taken to remove the spent substrate from the farm.

Ginger blotch

The ginger blotch disease in button mushroom is caused by *Pseudomonas gingeri*. The infected mushroom shows small, pale-yellowish-brown flecks on young sporocarps which develop into a reddish-ginger like structure as the mushroom matures (Wong et al. 1982). This colour variation readily distinguishes *P. gingeri* infections from *P. tolaasii* on the sporocarp. The pathogen also releases a toxin which results in superficial lyses of cap tissue. The lesions caused by *P. gingeri* remain superficial, culminating 1–2 mm below the mushroom cap surface, and are often confined to the circumference of the affected sporocarp. In later flushes, some cracking of the mushroom cap surface may become evident, but cap distortion or sunken lesions are not formed. As regards the taxonomy, the bacterium, *P. gingeri* is ascribed to the complex of *P. fluorescens*, but remains distinct from *P. tolaasii* (Nott 1989).

Management: The management practices are same as those of brown blotch disease. Maintenance of strict hygiene and sanitation in mushroom cultivation is required. Use of properly pasteurized compost and casing soil and avoidance of excessive humidity, high temperature and poor ventilation in growing room are effective measures to prevent the disease. Further, precaution is to be taken to avoid fluctuation of temperatures in the growing room which may cause water condensation on mushroom caps. Removal of diseased mushrooms which can be a source of the disease is to be done. Care should also be taken to control mushroom flies and mites which can spread the bacteria. Spraying of chlorinated water (125 ml of 10 per cent chlorine solution in 100 litre of water per 100 m^2 of bed area), streptomycin (0.02 per cent), oxytetracycline (0.03 per cent) or agricultural antibiotics, like pushamycin, plantamycin, agrimycin, etc. at 0.05 per cent of commercial formulations or application of terramycin at 9 mg/sq. ft. before first break, i.e. when the pin size is 4–5 mm, are effective to control the disease.

Mummy disease

The disease is caused by *Pseudomonas aeruginosa* which generally prevails in U.K. (Wuest and Zarkower 1991). Its occurrence in India is yet to be reported. It visibly appears in mushroom bed when the pin heads are developed. During that time, the pin heads appear in patches and several of them do not grow further and remain as stuck with the casing soil. The affected fruit bodies turn grayish in colour and open prematurely. In second flush, the stipe becomes crooked and caps are tilted. The base of the stipe becomes thick and is surrounded with fluffy mycelia. At last, the mushroom turns to tough, leathery dry and brown, appearing as mummy.

Management: Maintenance of strict hygiene and sanitation in mushroom cultivation is needed. Use of properly prepared substrate and casing soil that have been adequately pasteurized is essential. If the disease appeared in mushroom bed, isolation of the diseased areas by preparing ditches around and treating the infected region with formalin (0.5 per cent) are effective measures to check the disease severity and spread.

Drippy gill

The disease is most severe in autumn and winter months. Gills of sporocarp are attacked by the bacterium (*Pseudomonas agarici*) before the veil of the mushroom is broken. The affected gills are often underdeveloped, showing small, brown and decayed areas with profuse creamy bacterial ooze coming out from them. Discrete droplets of cream-coloured bacterial ooze coalesce in some cases with the droplets of adjacent gill to form ribbons of slime (Young 1970). Areas of gill in advanced stages of necrosis are distinguished by having either a light brown crusty appearance or a moist, black mucoid and adjacent gills are often coalesced. Longitudinal splits are evident in the stipes that often lined with bacterial ooze. In severe cases, the splits run along the length of stipe and pervade the stipe to such an extent that the inner core tissues are exposed (Gill 1994). The nature of symptoms, such as, expression of symptoms before veil break, suggests that the causal bacterium may subsist and indeed be transmitted intrahyphally (Rainey and Cole 1988, Fletcher et al. 1989). It is also suggested that the bacterium produces an enzyme which degrades the extracellular matrix, but does not elicit a detrimental response from the host, resulting in penetration of the sporocarp and reduction of stipe and hymenial integrity (Gill 1995).

Management: Use of disease free spawn and maintenance of hygiene in mushroom cultivation are effective measures to manage the disease.

Pit disease

The disease is common, but does cause any severe damage. Small dark and slimy pits of several mm deep and 1–10 in numbers appear on stipe and cap surface. The cause is not fully established. Different bacteria, viz. *Pseudomonas fluorescens*, *Erwinia carotovora* and *Bacillus polymyxa,* have been isolated in different investigations from the pitted region and supposed as the cause of the disease either solely or partially in association with mites (Wood 1950, Lelliott and Stead 1987, Richardson 1993, Gill 1995).

Management: Maintenance of hygiene in mushroom cultivation is needed. Spraying of chlorinated water (125 ml of 10 per cent chlorine solution in 100 litre of water per 100 m^2) or application of terramycin at 9 mg/sq. ft., streptomycin (0.02 per cent), oxytetracycline (0.03 per cent) or agricultural antibiotics, like pushamycin, plantamycin, agrimycin, etc. at 0.05 per cent of commercial formulations, as spray may be effective to control the disease.

Viral Disease

A large number of viruses or virus-like particles are found to cause diseases in button mushroom, particularly in *Agaricus bisporus*. The viruses, like MV-1 (spherical particles with diameter 25 nm), MV-2 (spherical particles with diameter 29 nm), MV-3 (bacilliform particles with size 18 × 50 nm), MV-4 (spherical particles with diameter 35 nm), MV-5 (spherical particles with diameter 50 nm), club shaped virus (with diameter 60–70 nm or 120–170 nm, spherical body of 40–50 nm and a cylindrical tail of 20–30 nm in diameter) and rod shaped viruses (with sizes 19 × 190 nm, 19 × 350 nm, 20 × 130 nm) are found to be associated with the diseases of mushroom (Sharma 1995). However, in India only two viruses, like MV-2 (spherical particles with diameter 29 nm) and MV-4 (spherical particles with diameter 35 nm) have been reported to cause diseases (Raychaudhury 1978, Tewari and Singh 1984, Goltapeh and Kapoor 1990).

Several names, such as 'La France disease', 'brown disease', 'watery stipe disease', 'X-disease' and 'die back disease' were used earlier to distinguish the viral diseases. Various types of symptoms developed due to virus attacks. In most cases, the diseased mycelium does not permeate or hardly permeates the casing layer or disappears after normal spread. Mushrooms appear only in dense clusters and become mature early. Mycelia which permeate grow slowly and develop into abnormal fruit bodies. Pin heads are slow growing and dwarf. Sporophores appear in late with less in number. The fruit body matures early and are off white in colour. The diseased mushrooms are loosely attached with the substrate. Sometimes, the stems of sporophores are elongated and the caps are small. Stipes become spongy and quickly turn brown on cutting and show an abnormal structure. The elongated stipes are bent. Stipes are watery and the streaking in stipes is visible. The veil opens prematurely. In certain cases veils are abnormal or absent. Gills are hard. A specific musty smell can be perceived in a growing room with diseased mushrooms. The virus transmits through mycelium, spores, vectors (phorid fly) or mechanically if purified virus hypodermically injected (Schisler et al. 1963, Hollings et al. 1963, 1971, Gandy and Hollings 1962, Van Zaayen and Temmink 1968).

Management: Maintenance of strict hygiene in every step of mushroom cultivation is essential. Use of disease free spawn is necessary. In order to free the virus from the culture, the virus infected culture has to grow in petriplate medium and incubate at 33ºC for two weeks. Later, the mycelia are cut from the advancing zone and subcultured at 25ºC. The newly grown culture is generally free from the virus (Gandy and Hollings 1962). Further, it is to mention that the pathogenic viruses of *A. bisporus* are not able to cause diseases in *A. bitorquis* (Van Zaayen 1976). So, in case of virus affected area the latter species may be cultivated. Another method to combat the viral disease as practiced in certain cases is the use of hybrid mushroom for commercial growing. The hybrid is prepared by hyphal anastomosing between the common white button mushroom and the button mushroom with brown, cream or off-white sporophores. The latter one is tolerant to viral diseases without showing marked symptoms and its wide spread culture reduces the effectiveness of virus (Fletcher et al. 1986, Romaine and Schlagnhaufer 1989).

DISEASES OF OYSTER MUSHROOM

Fungal Diseases

Four different types of fungal diseases caused by different fungal genera are common in oyster mushrooms.

Cladobotryum soft rot (cobweb)

Three species of *Cladobotryum*, viz., *C. apiculatum*, *C. verticillatum* and *C. variospermum* are found to develop as white cottony growth on substrate. They cause small brown irregular sunken spots on sporophore or fluffy growth on fruit bodies. Eventually, the fruit bodies decay with soft rot and emit foul smell (Sohi and Upadhyay 1986, Upadhyay et al. 1987, Goltapeh et al. 1989).

Management: Spraying of carbendazim (0.05 per cent), proper aeration in mushroom house and avoidance of excess watering are effective measures to control the soft rot disease.

Gilocladium brown rot

The fungi *Gliocladium virens* and *G. deliquescens* are found to cause spot disease in oyster mushroom. At first the mycelia of the causal fungus cover the young fruit bodies and develop green spots on sporophores. The affected young pin heads become soft, brown, pale yellow and decayed. Mature fruit bodies, while affected, show brown spots with yellowish halo (Bharadwaj et al. 1987, Sharma and Jandaik 1987).

Management: Spraying of carbendazim (0.05–0.10 per cent), proper aeration in mushroom house and avoidance of excess watering are the best measures to control the brown rot disease.

Arthrobotrys yellow rot

The disease is caused by *Arthrobotrys pleuroti*. The causal fungus develops as fluffy mycelial growth on substrate. The infected tissues of the fruit bodies become yellow waterlogged and rot (Ganeshan 1987).

Management: Spraying of carbendazim (0.05 per cent), proper aeration in mushroom house and avoidance of excess watering are effective measures to control the yellow rot disease.

Sibirina rot

The causal fungus, *Sibirina fungicola*, causes powdery white growth on fruit bodies. Subsequently, the infected primordia become brownish and rotten, and the mature fruit bodies become fragile (Jandaik and Sharma 1983, Sharma and Jandaik 1983).

Management: Spraying of benomyl (0.10 per cent), proper aeration and R.H. in mushroom house and avoidance of excess watering are the control measures of the rot disease.

Weed Fungi

A large number of fungi occur as competitor mould in oyster mushroom bed, particularly, when the pasteurization or disinfection of the substrate not done properly. In certain cases, unfavourable environmental conditions, high moisture and nutritional misbalance augment their appearances. These moulds cause crop losses up to 70 per cent. Further, being competitive in nature, they produce metabolites which directly inhibit the growth of mushroom mycelia. These weed fungi are as follows:

Arthrobotrys sp.

The fungus appears on substrate as fluffy growth of mycelia. The conidiophores are long, slender, hyaline and slightly enlarged at the apex. Conidia are ovate-oblong with two unequal celled and borne in cluster.

Aspergillus sp.

Different species of *Aspergillus* cause various types of moulds on a mushroom bed, like *Aspergillus niger* produces black mould, while, *A. flavus* produces yellow mould and *A. fumigatus* produces grayish green mould.

Alternaria alternata

The fungus produces dark conidia with both transverse and longitudinal septa. The obclavate condia are born acropetally in chains on dark conidiophore.

Cephalosporium sp.

Different species of *Cephalosporium*, viz. *C. aspermum* and *C. acremonium*, are found to appear in oyster mushroom beds. The fungi produce simple, hyaline, one celled conidia (phialospores), which are collected in slime drop on slender phialides.

Chaetomium globosum

The fungus produces olive green mould. It forms perithecia which are scattered or gregarious, ovate or ellipsoid, often pointed at the base and, thickly and evenly clothed with slender hairs. Inside perithecia, there are oblong-clavate and evanescent asci. The ascospores are broadly ovoid and faintly apiculate at both ends.

Cladosporium cladosporoides

The fungus produces dark condia (blastospores) of 1–2 celled on tall and dark conidiophore which is variously branched near apex.

Coprinus sp.

Several species of *Coprinus*, viz. *C. atramentarius*, *C. lagopus*, *C. fimitarius, C. retirugis, C. sterguilinus*, etc. are found to grow on mushroom beds. The weed fungi appear as slender bell shaped mushrooms with scales on caps which on maturation decay and form blackish slimy mass due to auto digestion.

Cochliobolus specifer

It forms conidiophores solitary or in small groups. The conidiophores are flexuous, repeatedly geniculate with numerous well defined scars and brown in colour. Conidia straight or oblong or cylindrical, rounded at the ends, when mature golden brown except for a small area just above the dark scar, smooth, with 3 pseudoseptation. The perfect state forms black ellipsoid to globose perithecia with well defined ostiolate beak.

Drechslera bicolor

The fungus produces brown and simple conidiophores. The conidia (porospores) are dark, several celled, cylindrical and borne at the apex of conidiophores through small pores. The conidiophore grows sympodially from the point below apex and then forms a second spore on a new apex.

Fusarium sp.

Different species of *Fusarium*, such as *F. moniliforme and F. graminearum*, are found to be appeared in mushroom beds. These fungi produce two types of spores (macrospores and microspores). The macrospores are large fusiform and slightly curved with several cells, while, microspores are small with 1–2 cells.

Mucor sp.

The fungus causes bread mould. The aseptate sporangiophore bears black spores in a sporangium which have columela.

Neurospora sitophila

The fungus causes pink mould (Dayaram et al. 2008). It appears as whitish to pinkish and yellowish mycelial growth in bag during spawn run condition within 5–7 days of spawning. Thereafter, within 2–3 days it covers the whole bag and retards the growth of mushroom mycelia. Small yellow fruit bodies of the mould also appear at later stage.

Penicillium sp.

The fungus is found to cause blue-green mould on mushroom beds. It produces chains of spores on finger like conidiophores.

Rhizopus sp.

Rhizopus oryzae is the most common mould on the mushroom beds prepared with paddy straw. Several other species of this genus are also found to appear in mushroom beds. The fungi appear as black mould with dark spores in sporangium that has no columela.

Stachybotrys chartarum

The mould produces simple conidiophores bearing at apex a cluster of globose to ovoid dark conidia on thick and short phialides.

Stilbum nanum

The mould grows as with synnemata bearing one celled globose condia at its head on slender vertically branched conidiophores.

Stysanus medius

The hyphae and conidiophores of the mould are dark. The conidiophores are compacted into synnemata with dense head of sporogenous cells and chains of conidia, or may remain as solitary. Upper part of the conidiophores is branched penicilliately, producing masses of dry spores epically on sporogenous cells. Conidia are mostly ovoid, 1-celled and dark.

Sclerotium rolfsii

The fungus produces dark brown globular sclerotia which are asexual fruit bodies.

Sordaria fimicola

The mould forms perithecium with hairy neck. The spores are continuous and dark coloured.

Oedocephalum sp.

Different species of *Oedocephalum*, such as *O. globerulosum* and *O. lineatum,* are found to occur in mushroom beds as mould. The mould produces simple and hyaline conidiophores which are enlarged and globose at apex. At the head of the conidiophores there are many dry, hyaline, 1-celled and globose to ovoid condia. The spore bearing portion is denticulate.

Trichoderma viride

This fungus causes green mould on mushroom bed. Conidiophores are hyaline, much branched, not verticillate, phialides are in groups. Conidia (phialospores) are hyaline, 1-celled, ovoid, rough walled and borne in small terminal clusters. The mould is usually recognized by its rapid growth at 20ºC and slow growth at 27ºC, as well as by the green patches or cushions of conidia.

Trichoderma harzianum

This is very similar to *T. viride* in causing green mould. It appears in humid condition during June and July most abundantly in Assam (Siddique et al. 2004).

Trichothecium roseum

The mould produces simple, long, slender and septate conidiophores. The conidium borns singly or successively at the apex of conidiophore that grown slightly after each conidium formation. The conidia are ellipsoid, 2-celled and hyaline or brightly coloured.

Trichurus terrophilus

The mould forms dark synnemata with slender stalk; expanded spore bearing portion; long, black, simple or branched hairs or spines among the conidiophores and dark, 1-celled, ovoid, catenulate conidia.

Phialospora sp.

This mould produces conidiophores which are short or reduced to phialides, dark, simple or branched; phialides are cylindrical to inflated, often with flaring collaret at apex. Conidia (phialospores) are sub hyaline to dark, 1-celled, globose to ovoid and extruding from phialides in moist heads.

Management: The severity of these mould attacks can be well controlled by spraying carbendazim (0.05 per cent) just after opening the polythene bag on completion of spawn run (alternatively, by applying benomyl 50 ppm, carbendazim 100 ppm + blitox 100 ppm or thiram 100 ppm). (Bano et al. 1975, Sharma and Jandaik 1980, 1981, Doshi and Singh 1985, Singh and Saxena 1987, Vijay and Sohi 1989, Das and Suharban 1991).

Bacterial Diseases

A number of bacteria are found to cause diseases in oyster mushrooms (Fermor 1987, Shymanski 1991). In most cases, different species of *Pseudomonas* cause diseases of oyster mushrooms. The diseases appear as brown depressed spots or yellowish blotch and eventually, rotting with foul smell. In case of early attack, particularly at primordial stage, entire flush shows rotting. The causal organisms also develop brown patches on spawn run substrate due to spoiling or retarding of mycelial growth. The major diseases are as follows:

Bacterial distortion

This disease is caused by *Pseudomonas fluorescens*. It causes distortion of fruit bodies that appeared as fist-shaped fruit bodies.

Bacterial spot

Bacterial brown spots appear on fruit bodies due to infection of *Pseudomonas stutzeri* (Mallesha and Shetty 1988).

Bacterial rot

The fruit bodies show water soaked areas and yellow brown discolouration. Rotting of grown up fruit bodies starts at the centre and extends towards the periphery. The gills and lower surface turn yellow and the caps become crinkled and rolled upwards and inwards. The disease is caused by *Pseudomonas alcaligens* which gives whitish effuse growth on PDA medium (Biswas et al. 1983).

Brown blotch

Brown blotch is very common disease all over the world. It is caused by *Pseudomonas tolaasii.* It causes various types of blotching and mottling of fruit bodies.

Yellow blotch

This disease causes heavy loss in India (Jandaik et al. 1993). The causal organism is *Pseudomonas agaraci* (Besstte et al. 1985). It appears as blotches of varying sizes on pilei as yellow, hazel brown, fawn or orange in colour. If the infection is taken place at early stage the infected fruit bodies turn yellow and remain stunted.

Management: The most effective measure to control bacterial diseases is spraying of streptomycin sulphate (0.025 per cent) or equal concentration of streptomycin and tetracycline hydrochloride mixture which commercially sold as *Plantamycin, Pushamycin, Agrimycin,* etc. by a hand sprayer on the spawn run substrate after opening the polythene bag or polythene sheet or as and when the disease attacked. Proper disinfection or pasteurization of substrate, cleaning of racks and floor with 5 per cent bleaching powder solution and maintenance of hygiene are very effective measures to prevent the disease attack.

Viral (Virion) Disease

Virions measuring 26 ± 2 nm and 21 nm in diameter have been found to affect oyster mushroom (Liu and Liang 1986, Liang et al. 1987, Sharma 1995). The disease causes curling of pileus upwards, swollen or elongated stalk, premature spore shedding and very poor yield.

Management: The disease is controlled by growing the culture on a desired medium in petriplates at 33°C for two weeks. The mycelia grown in that higher temperature are cut out from the advancing zone with the aid of a cork borer and placed aseptically in a slant medium. The culture is now free from disease and allowed to grow at 25°C.

DISEASES OF PADDY STRAW MUSHROOM

Paddy straw mushroom is affected by the diseases, like wet bubble (*Mycogone perniciosa*) and button rot (bacteria, *Pseudomonas* sp.). However, the existence of the former disease in India is yet to be established, although, in the major paddy straw mushroom growing countries, like China, Taiwan, etc. the wet bubble disease is the very destructive one (Chang 1978a). Apart from the diseases, several weed fungi are found to grow over the bed. For example, the white plaster mould *Scopulariopsis fumicola* and a very similar fungus *Verticillium agaricicotum* are common moulds of paddy straw mushroom. In case of cotton waste compost using as substrate, *Thielavia terricola*, *Trichoderma* sp., *Aspergillus* sp., *Pythium* sp. and *Rhizoctonia* sp. are the common occurrence in paddy straw mushroom cultivation. In India, the moulds, like *Chaetomium* sp., *Alternaria* sp., and *Sordaria* sp. are observed when wheat, barley, jowar, etc. are used as substrates. In case of paddy straw substrate, several *Coprinus* species, like *C. aratus*, *C. cinereus*, *C. lagopus*, etc. and various other moulds, such as *Psathyrella* sp., *Penicillium* sp., *Podospora favrelii*, *Aspergillus* sp., *Rhizoctonia solani*, *Rhizopus* sp. and *Sclerotium* sp. are the common weeds in paddy straw mushroom cultivation in India (Kannaiyan 1974, Munjal 1975, Chang 1978, Bahl 1984, Purkayastha and Das 1991, Rangaswami 1978, Bahl and Chowdhury 1980). The damage caused by *Coprinus* sp. is the greatest problem in paddy straw mushrooms. The mould completes its life cycle in shorter duration (1 week) than the straw mushrooms.

Management: The best measure to obtain good yield with minimum growth of *Coprinus* is by keeping C: N ratio of the substrate in the range of 40:1 to 50:1. The moisture content of the substrate should be kept in range of 60 to 65 per cent, since the high moisture favours the growth of *Coprinus*. Several additional methods are being used to control the diseases and moulds. The pasteurization is one of the most effective measures to control them. Partial disinfection of straws by dipping the straw bundles in carbendazim (75 ppm) and formalin (500 ppm) mixed solution for 10 minutes before bed preparation is also used to minimize the diseases and moulds (Biswas and Singh 2008). Partial sterilization of straw and spraying of zineb (0.2 per cent) or captan (0.2 per cent) are also suggested to manage the diseases and moulds in paddy straw mushroom cultivation. Further, the spraying of insecticide, fungicide and antibiotics (Malathion 0.025 per cent + mancozeb or benomyl 0.025 per cent + tetracycline 0.025 per cent) in combination is the recommended measure to manage both disease and pest, simultaneously, in paddy straw mushroom cultivation (Kannaiyan and Prasad 1978).

DISEASES OF OTHER MUSHROOMS

Fungal Diseases

The giant mushroom (*Stropharia rugoso-annulata*) is found to be affected by *Mycogone rosea*. The pathogen causes white cottony growth on gills, light brown spots on stipe and deformity of sporophores (Sohi and Upadhyay 1989). The *Jew's* ear mushroom (*Auricularia polytricha*) is quite often affected by cobweb disease caused by *Cladobotryum verticillatum*. The causal fungus causes white fluffy growth on substrate and fruit bodies (Goltapeh et al. 1989). Several competitor moulds, like *Aspergillus niger*, *A. flavus*, *A. fumigatus*, *Rhizopus stolonifer*, *Mucor* sp., *Sclerotium rolfsii*, *Trichoderma viride*, *T. haematum*, *Fusarium* sp. and *Coprinus* sp. have been found to grow in milky mushroom (*Calocybe indica*) bed (Doshi et al. 1991). Further, the moulds *Cadobotryum* sp. is found to appear in high humid condition on casing layer as the growth of small circular white patches of mycelia. Another mould, *Oedocephalum* sp. forms irregular, light silver gray to dark tan or light brown patches on surface of casing layer. The Shiitake mushroom is commonly affected *Trichoderma viride* (Thakur and Sharma 1992).

Management: The above mentioned mushrooms are cultivated either on pasteurized or sterilized substrates. Special care is needed to do both the operations correctly. Further, if moulds appeared, spraying of carbendazim (0.05 per cent) (differently, 50 ppm) is effective to control the moulds and diseases.

Bacterial Diseases

The winter mushroom *Flammulina velutipes* is found to be affected by two bacterial diseases. Brown soft rot disease caused by *Erwinia* sp. and disorder of fruit bodies by *Pseudomonas tolaasii* are frequently observed when the mushrooms are grown in saw dust medium (Fermor 1987, Suyama and Fujii 1993). The pathogen secretes a toxin which results in superficial lyses of cap tissue. In *Lentinula edodes* tissue browning and fruit body distortion are occurred due to the infection of *Pseudomonas fluorescens* (Komatsu and Goto 1974). The bacterium, *P. tolaasii* has been found to develop range of symptoms from mid browning to severe necrosis and cavity formation (Tsuneda et al. 1995). In case of Jew's ear mushroom (*Auricularia polytricha*) bacterial disease is rare, although several bacteria have been found commonly as secondary invader following fungal attack (Gill 1995). A single bacterial disease caused by an unknown bacterium is reported (Fermor 1987). The milky mushroom is found to be affected by brown blotch disease caused by *Pseudomonas* sp.

Management: It is necessary to maintain strict hygiene and sanitation in mushroom cultivation. Use of properly disinfected or adequately pasteurized substrate is needed. Avoidance of excessive humidity, high temperatures and poor ventilation in growing room is essential.

Care should also be taken to control mushroom flies and mites which can spread the bacteria. Spraying of chlorinated water (125 ml of 10 per cent chlorine solution in 100 litre of water per 100 square metre bed) or agricultural antibiotics, like pushamycin, plantamycin, agrimycin, etc. which are the mixture of streptomycin sulphate and tetracycline hydrochloride, at 0.05 per cent of commercial formulations are effective to control the disease. Further, precaution should be taken to remove the spent substrate from the farm.

PESTS

Mushrooms are attacked by different types of insects, mites, fungal eater nematodes, etc. However, amongst them, sciarid and phorid flies, springtails and mites are the important arthropod pests of cultivated mushroom in India. The flies lay their eggs in the substrate during spawn running stage. The larvae (grubs) come out from the eggs, eat mycelia and form tunnels in mushroom sporophore, which become soft and rot with brownish tinge. The adults mostly eat spores or suck the juice from the soft part of gills and occur gregariously on ventral surface of cap. The important pests are as follows:

Sciarid Flies (Diptera: Sciaridae)

The sciarid flies, also called as 'dark-winged fungus gnats', generally occur in moist shady places and the larvae live in fungi. Several species of sciarid flies, viz. *Bradysia paupera*, *B. tritici* and *Lycoriella auripila*, have been found to damage mushrooms in India (Shandilya

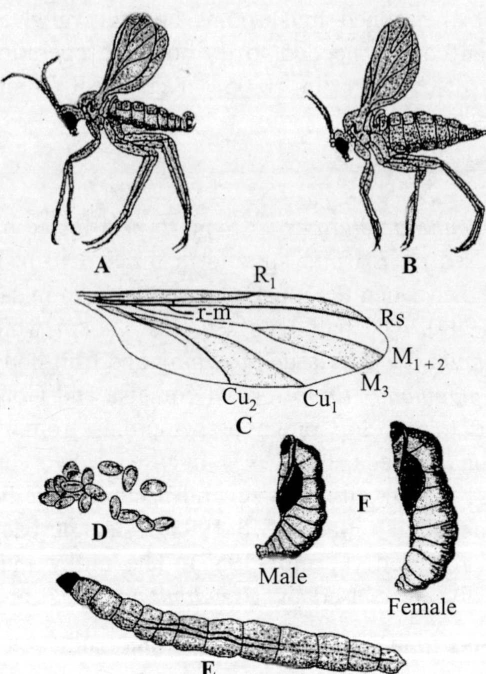

Figure 8.5 Sciarid flies (*Bradysia tritici*). **A.** Male adult, **B.** Female adult, **C.** Wing with characteristic venation, **D.** Eggs, **E.** Larva and **F.** Pupae.

et al. 1975, Sandhu and Brar 1980, Chakravarty et al. 1987). The sciarids are grayish black winged and 2.3–3.2 mm long flies having long thread like antennae with several segmentation and the eyes meeting above the bases of the antennae (Figure 8.5). Radial and radial sector veins in wings are thick and the latter one is unbranched with a cross-vein in between. The r–m cross vein is in the line with and appears as a basal extension of RS. The adult lays about 110 eggs in clusters. The freshly laid eggs are pulpy white, round to oval in shape. The larvae are with distinct shining black heads and dirty white bodies which are transparent and show the presence alimentary canals from out side. The newly formed pupa is dirty white and at maturity it becomes grey to black. The sciarid flies lay their eggs in the substrate or compost. The emergent larvae feed on both mycelia at the spawn running stage and fruit body during cropping period by entering into the pin heads through prepared holes. The affected fruit body becomes spongy, hollow and brown in colour. The growth of the infested fruit bodies is stopped and that turns into brown in colour, leathery and unsuitable for consumption. However, when the fruit bodies of button mushrooms are attacked at later stage the damage mainly confines at the stipe and the cap remains free from attack, which on cutting from the stipe is usable for food (Brar and Sandhu 1990). However, in oyster mushroom both stipes and caps of larger sporophores are extensively tunnelled and contaminated with faeces of the flies (Chakravarty et al. 1987).

Management: The flies are generally controlled following the integrated pest management system in order to minimize the pest.

Sanitary measure: Maintenance of hygienic condition is very good preventive measure for pest attacks in general. Proper sanitation measure is to be taken to clean the mushroom house and its surroundings. The spent compost, casing materials and substrates are needed to put in a compost pit and that has to cover with a layer of manure or soil up to 10 cm of thickness. This will check the breeding and multiplication of flies at the surrounding area.

Screening measure: Fixing of nylon net (14–16 mesh/cm) at doors, windows and ventilators is effective to screen out fly's entry inside the mushroom house (Sandhu and Arora 1990). In case of oyster mushroom, one unique method is used to restrict flies attack during spawn running stage. In this method, instead of perforated polybag, intact poly bag is filled with substrate and spawn. The bag mouth is closed with a thread and mustard oil is pasted all over the bag. Now, perforation is made at different places with sharp stick and cotton is inserted at the perforated holes. By this way the fly's entry can be avoided.

Chemical measure: Apart from the above two methods, certain prophylactic measures are taken to prevent pest attack. It is suggested to mix lindane thoroughly with the substrate mixture during last turning of compost making @ 20 ml of lindane 20 EC/100 kg of straw (Gill and Mangat 1991). In case of substrate soaking in oyster mushroom cultivation, addition of 5 drops endosulfan 35 EC in 20 litres of water is effective to check the flies attack. Simultaneously, this measure prevents rat attacks in spawned substrate, which often occurred during spawn running stage. In case, the flies had already attacked the mushrooms, then spraying of malathion (0.01 per cent) on mushroom beds after harvesting all the growing fruit bodies and in air inside is very effective to minimize the pest during further flushes (Biswas and Singh 2008b).

Phorid Flies (Diptera: Phoridae)

The phorids are called 'humpbacked flies'. These are small flies that easily recognized by the humpbacked appearance, the characteristic venation and laterally flattened hind fimora (Figure 8.6). The species, *Megaselia agarici* (Linter) (= *Megaselia sandhui* Disney) has been found to damage mushrooms in India. The phorid flies are light to dark brown in colour and 1.9–2.0 mm long in size. The wings are rounded apically with anterior strong veins and posterior weak vein. Antennae are 1-segmented with long aristae and the hind legs are long whose femora are flattened laterally. The eggs are whitish, elongated cylindrical and slightly curved. The larvae are dirty white with narrow transparent anterior end and visible black mouth hooks. The pupae are dorsiventrally flattened and light to dark brown in colour (Sandhu and Bhattal 1986). The larvae emerging from eggs feed on mushroom mycelia and tissues. In fruit bodies they form tunnels and move upwards. Attack at the pinning stage restricts further development of pinheads and buttons. The flies attack mostly during October–November.

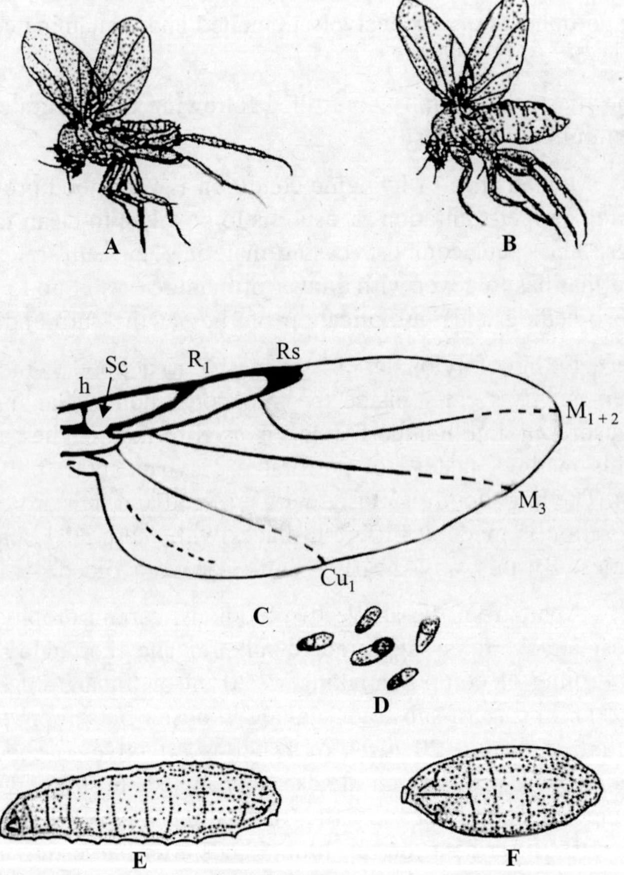

Figure 8.6 Phorid flies (*Megasella agarici*). **A.** Male adult, **B.** Female adult, **C.** Wing with characteristic venation, **D.** Eggs, **E.** Larva and **F.** Pupae.

Management: Management of phorid flies is similar to that of sciarid flies. Here also, maintenance of hygienic condition is very much essential to prevent the pest attacks. The entry of the flies is checked by putting nylon nets (14–16 mesh/cm) on windows and doors. The addition of traces of endosulfan 35 EC (@ 5 drops/20l solution) during substrate soaking or disinfection in chemical method is very effective measure to control the fly's attack during spawn growth in substrate. Simultaneously, this measure also prevents rat attacks in spawned substrate (Biswas and Singh 2008b). Alternatively, spraying of dichlorovos (0.01 per cent) on mushroom beds and in air inside after 7 days of spawning and spraying 2nd time after casing (in case of button mushroom) or opening of bags (oyster mushroom) may be suggested (Sandhu 1995). Further, if infestation has already been established, then spraying of malathion (0.01 per cent) after harvesting of all the growing fruit bodies is very effective to control the fly's attack during subsequent flushes.

Springtails (Collembola)

Different springtail insects, viz. *Lepidocytrus cyaneus, Seira iricolor* and *Xenylla* sp., have been found to damage mushrooms at different places of India, like Himachal Pradesh, Delhi, Rajasthan and Punjab (Shandilya et al. 1975, Bahl et al. 1981, Thapa and Seth 1983, Bhandari and Singh 1983, Gill and Sandhu 1994). The springtails are minute insects (2.85 mm in case of *S. iricolor* adult). They are so named because they have forked structure or furculum with which they jump. The furculum arises on the ventral side of the fourth abdominal segment. During rest period the furculum is folded forward under the abdomen. The adults are of ground colour with light violet band alongsides of the body. Intensely dark and rounded scales are present all over the body. The mouth parts are elongate and stylet like and are concealed within the head. Antennae are short and with a few segments (Figure 8.7). The springtails feed on mycelia in the spawned compost or substrates. In case of button mushroom, the insects attack fruit bodies and cause slight pitting and browning at feeding sites. In oyster mushroom, the springtails feed on gills resulting in destruction gill linings and formation of webby structures. The springtails live in moist soil and organic matter. They have no wings and their entry in the mushroom bed is with the contaminated substrates or compost (Sandhu 1995).

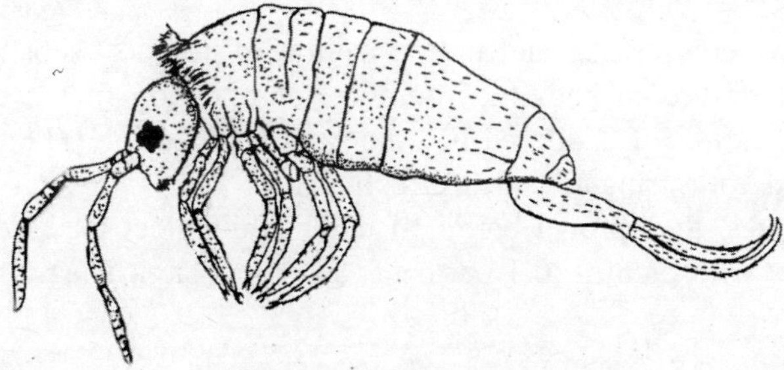

Figure 8.7 Springtails.

Management: Floors of compost yard, mushroom house and the surroundings are kept as cleaned properly. Thereafter, the floors are disinfected by spraying malathion (0.05 per cent) emulsion. Using of properly pasteurized compost (for button mushroom) or substrates (for oyster mushroom) is effective measure to control the pest. Further, in case of unpasteurized compost, incorporation of diazinon in the compost at 30 ppm dosage (15 ml of diazinon 20 EC diluted in water and sprayed in 100 kg compost) at filling time is recommended. The infestation during spawn run or cropping period can be checked by spraying malathion (0.05 per cent) or dichlorovos (0.025 per cent) from time to time.

Mites (Acari)

Although a large number of mites are present in mushroom beds, only a few are the harmful pests of mushroom. The saprophagous mite, viz. *Tyrophagus putrescentiae*, is the most damaging species of mite occurring in Himachal Pradesh, Delhi, West Bengal and others (Bahl et al. 1981, Thapa and Seth 1982). The mites of this species are 0.25 mm long and yellowish-brown in colour and with four appendages (Figure 8.8). The mites feed on mycelium and damage sporophores by causing shrunk caps and brown rusted spots on buttons. They may completely destroy the young buttons of mushrooms, while attacked at early stage of development. The mites enter into mushroom bed by flies, on which the migratory stages of mites are clung by means of suckers.

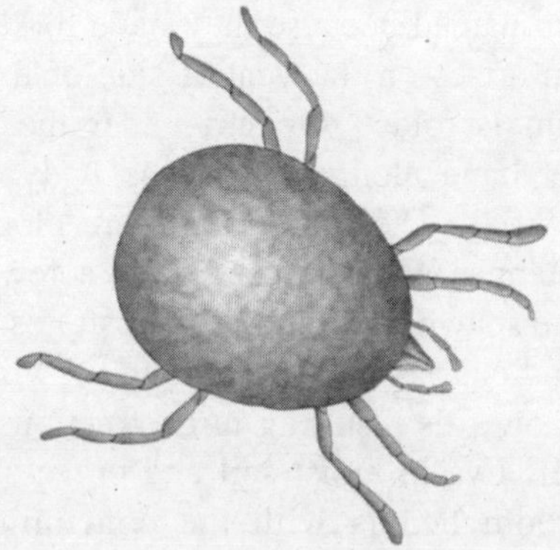

Figure 8.8 Mite.

Management: Proper composting and pasteurization of compost/substrate are effective to avoid mite attack in mushroom bed. Incorporation of diazinon in the compost at 30 ppm dosage (15 ml of diazinon 20 EC diluted in water and sprayed in 100 kg compost) at filling time as recommended for springtails also effective to control mites. If a given mushroom crop is affected by mites, then proper care is needed to disinfect the floors of compost yard and mushroom houses with dicofol (0.1 per cent) as prophylactic measure before subsequent crops.

CHAPTER 9

Post Harvest Technology and Food Preparations

The production and consumption of mushrooms are increasing very fast all over the world. At present, about twenty species of mushrooms are being cultivated commercially, of which, a few are produced in industries. Significant production of mushrooms in the world occurs of button mushroom (*Agaricus bisporus*), shiitake mushroom (*Lentinula edodes*), oyster mushroom (*Pleurotus* sp.), black ear mushroom (*Auricularia polytricha*) and paddy straw mushroom (*Volvariella volvacea*). In India, the button mushroom which contributes about 85 per cent of the total mushroom production is grown throughout the year in the environmentally controlled export-oriented units. Besides, it has also been produced by a large number of seasonal growers of different provinces during the winter season, December to February. In addition, oyster mushroom, paddy straw mushroom and milky mushroom (*Calocybe indica*) are grown commercially by a large number of seasonal growers. Post harvest losses are very high in mushrooms, since, mushrooms even after harvesting continue to grow, respire, mature and senesce resulting in weight loss, veil opening, browning, wilting and finally in spoilage. Further mushrooms are highly perishable materials and get spoiled due to liquefaction, loss of texture, aroma, flavour, etc. making them unsuitable for sale (Azad et al. 1987). Most of the mushrooms, being high in moisture and delicate in texture, cannot be stored for more than 24 hrs at the ambient conditions that prevailed in the tropics.

India is predominantly a market for fresh mushrooms. Thus, during peak production period (mostly during winter), the supply of mushrooms at the local market exceeds the demand, causing less profit due to fall in price, and market surplus suffers spoilage. In most cases there are gluts in market and to overcome such gluts, different techniques for both short-term and long-term preservations of mushrooms are practiced (Figure 9.1).

Figure 9.1 Packed and preserved white button mushrooms. **A.** Short-term storage, **B.** long-term storage.

SHORT-TERM PRESERVATION

Washing

Washing is the first step for preserving mushroom. In Indian condition, due to use of soil in casing material, washing is required to remove adhered soil particles. However, in American and European countries, where peat is used as casing material, washing is not required. Simply, plain water may be used to remove adhered soil particles. However, the washed mushrooms generally deteriorate rapidly during storage due to presence of more water content than the mushrooms packed as dry. Further, washing with plain water does not improve any whiteness of the mushroom. Thus, to improve storing quality and maintain whiteness different pre-treatments have been suggested by various workers during their times.

Method 1: Pruthi et al. (1984) showed positive effect when mushrooms are dipped in dilute solution of hydrogen peroxide (1:3) for half an hour and subsequently, steeped in a solution containing citric acid (0.25 per cent) and sulphur dioxide (550 ppm) mixture.

Method 2: Use of sodium hypochlorite (100 ppm) and calcium chloride (0.55 per cent) with oxine (100 ppm) as pre-treatment has been found effective to increase antibacterial effects and lower cap opening during storage. In this solution, oxine, which is a stabilized form of chlorine dioxide is used to control bacterial growth, while calcium chloride to lower cap opening (Beelman 1987).

Method 3: Mushrooms may be washed in hard water (with 150 ppm calcium carbonate) to reduce both bacterial growth and colour deterioration during storage (Guthrie and Beelman 1989).

Method 4: Washing of mushrooms can be done in a solution consisting of oxine (50 ppm), sodium erythoborate (0.1 per cent), and calcium chloride (0.5 per cent). This solution is very effective to lower bacterial populations and colour deterioration during the storage (Guthrie and Beelman 1989).

Method 5: In other way, washing of mushrooms can also be done in 0.05 per cent potassium metabisulphite (KMS) solution for improving the initial whiteness which lasts longer period during storage (Saxena and Rai 1988).

Packaging

Storage of fresh mushrooms, i.e. short-term preservation, is done mostly for transportation and overcome the gluts of fresh mushrooms in markets. Before storage, packaging of mushrooms is prerequisite and plays very important role in handling, marketing and protecting the quality during storage and transportation. The decent packaging also increases the consumer's choice by seeing freshness of the mushroom through the packaging from out side. Generally see-through packaging is done in case of mushrooms and that increases consumer's confidence in the product.

Conventional Packaging

In case of button mushrooms, the harvested mushrooms are cleaned by cutting off the base by a sharp knife. The base-cut mushrooms are packed in small polythene or polypropylene packets containing 200 or 400 g mushrooms. Mushrooms are also packed in plastic punnets (trays) over wrapped with PVC (Polyvinyl Chloride) film. Generally, in traditional way button mushrooms are washed before packing in polypropylene (<100 gauge) bags. However, recently some farms have introduced to pack unwashed mushrooms in plastic trays over wrapped with PVC film. Six such trays each with 200 g mushrooms are placed in card board boxes for transportation. In this case non-perforated bags or PVC film is most suited for use (Saxena and Rai 1988). However, in oyster mushroom, the mushrooms are packed in perforated (with 5 per cent vent area) polypropylene bags after cutting stems and cleaning the adhered straws. The bags are stacked in trays or baskets. A few pouches of crushed ice are kept along with mushrooms for transport. The paddy straw mushrooms are packed at button stage in polythene bags with perforations as well as in tray packs. The mushrooms packed in bamboo baskets with an aeration channel at the centre and placed above dry ice wrapped in paper is in practice for transportation paddy straw mushrooms in Taiwan. While, in China, packing in wooden cases for transport by rail or boat is in practice (Saxena and Rai 1990). The packing of milky mushroom is more or less similar to button mushroom. The milky mushroom can be stored for 6 days in ambient condition and 9 days in refrigerator, while that packed in non-ventilated low density polyethylene bags (Sohliya et al. 2010).

Modified Atmosphere Packaging (MAP)

Modified Atmosphere Packaging (MAP) of mushrooms delays senescence and maintains quality of mushrooms during post harvest storage (Burton 1988, Henze 1989, Burton and

Maher 1991, Briones et al. 1992, Saray et al. 1994, Roy et al. 1995, Tano et al. 1999, Rai and Arumuganathan 2008). In this method a modified atmosphere is created in a sealed package of a fresh product by exchanging respiratory gases, like oxygen (O_2) intake and carbon dioxide (CO_2) evolution. It is evident that when the rate of gas permeation through the packaging material equals to the respiratory gas exchange, then an equilibrium of the concentrations of O_2 and CO_2 is established. This equilibrium depends on the temperature, respiration rate of specific product, product weight, O_2 and CO_2 permeability of the packaging material, free space in the package and film area (Bano et al. 1997). Modified atmosphere can be created by any of the following two ways.

Passive way: In passive way, modified atmosphere can be created by just putting the product in a sealed polymeric package. In such case, due to respiration of the fresh product and permeation of gases into the package, the atmosphere inside the packet becomes modified. In practice, the punnets (trays) are over-wrapped with differentially permeable Polyvinyl Chloride (PVC) or poly acetate films, and thus, created modified atmosphere of about 10 per cent CO_2 and 2 per cent O_2 within the trays during storage. This atmosphere is suited for keeping the quality of the mushrooms, since, it is established that the critical limits of CO_2 and O_2 for keeping best quality of mushrooms are 12 per cent and 1.5–2 per cent, respectively (Halachmy and Mannheim 1991). The shelf-life of button mushroom which covered with PVC-film is 5–7 days at 15–20°C as compared to 2–4 days shelf life of uncovered mushroom at same condition (Gormley and Mac Canna 1967). The combined effect of MAP and storage at low temperature increases shelf life further by delaying post harvest development of mushrooms. This packaging (MAP) system has an advantage that the mushroom can store at very low temperature (1.5°C) without any freezing injury (Halachmy and Mannheim 1991). Recently, antifogging film has been introduced for over-wrapping the punnets. The mushrooms covered with antifogging film can be stored up to 24 days (Chi et al. 1996, 1998).

Active way: This method is same as passive way, except, that in this case desired air is flushed into the package initially, so that the steady state of atmosphere is reached quickly after packaging. On the contrary, in passive way of atmospheric modification takes place after a long time of period to reach the steady state conditions of atmosphere within the package.

Controlled Atmosphere Packaging (CAP)

As compared to Modified Atmosphere Packaging (MAP), very less work has been done on Controlled Atmosphere Packaging (CAP). In this system, the quality of mushrooms is kept by incorporating desired atmosphere in the closed packet. In this system, non-perforated polythene packets (150 gauges or 300 gauges) are used to keep mushrooms. In case of oyster mushroom, 15 per cent CO_2 and 1 per cent O_2 are inserted inside the packets and sealed for storing at cold temperature. This system of packaging can keep mushroom quality up to 20 days (Ramanathan et al. 1992). However, in case of button mushroom, 10 per cent CO_2 and 8 per cent O_2 are the best compositions for the inside atmosphere of sealed packets (Zeng and Xi 1994).

Modified Humidity Packaging (MHP)

The polymeric films which are used in conventional packaging have lower water vapour transmission rates in comparison to that of fresh mushrooms. Thus, in most cases, there exists very high (nearly at saturated condition) relative humidity within the packets. This high relative humidity causes condensation of water vapours and allows microbial growth, which in turn decomposes mushrooms. The relative humidity of 87–90 per cent in the package is considered as the best suited condition for keeping the colour of mushrooms during storage (Rai and Arumuganathan 2008). Thus, to obtain the desired In-Package Relative Humidity (IPRH), the packets are either perforated to release the excess moisture outside or some water absorbing materials, like calcium chloride ($CaCl_2$) are kept inside to absorb IPRH (Eaves 1960). Moreover, it is evident that the combination of Modified Atmosphere Packaging (MAP) and Modified Humidity Packaging (MHP) further improves shelf-life of fresh mushrooms (Roy et al. 1996). Thus, the commercially available silca gel or montmorillonite clay may be used as moisture absorber along with differentially permeable PVC wrapped mushroom (as in MAP) during storage (Anantheswaran and Sunkara 1996).

Storage of Fresh Mushrooms

Cooling or refrigeration

Cold preservation is very essential aspect to store fresh mushroom for a certain period in order to transport and overcome the gluts in markets. In storage, the temperature of mushrooms rises steadily due to respiration and atmospheric temperature. This heat causes deterioration in quality and spoilage of mushrooms. Hence, the heat should be removed immediately after harvest and the temperature of mushrooms should be brought down to 4–5°C as soon as possible. The choice of the cooling system depends upon the quantity of mushroom to be handled, which may be a refrigerator for a small grower or a cold room with all facilities for a commercial grower. Using forced-chilled air, ice bank or vacuum cooling systems, mushrooms can be cooled up to 2–4°C. Storage under low temperature is an excellent method for restricting spoilage of harvested mushrooms for a certain period. The button mushrooms can be stored as fresh for 14–20 days at 1°C, while, for 10 days at 6°C and 2–3 days at 20°C storage (Minamide et al. 1980). As the quality of paddy straw mushroom deteriorates at low temperature storage, thus, it is stored at 10–15°C in perforated polyethylene bags. The size and shape of the packages also play certain important roles in cooling room. The packages with more than 10 kg mushrooms or above 15 cm thick layer of mushrooms cause problems in exchanging gases and cooling uniformly. In storage, vertical flowing of chilled air is given for cooling. This force-chilled air cooling system is time consuming, and thus, vacuum cooling is becoming more popular, nowadays.

Vacuum cooling

In vacuum cooling, the mushrooms are exposed to very low pressure. At this condition, the inside water of the cell walls and interhyphal spaces of mushrooms are evaporated giving off the latent heat of vapourization, thus, cooling themselves from ambient to 2°C in 15 to 20 minutes. This vacuum cooling is a uniform and faster process than force-chilled air cooling. Further, the vacuum-cooled mushrooms are superior in colour during storage than conventional ones. However, this cooling system has certain drawbacks. For example, this system requires high capital cost. It causes loss of fresh weight during the process of cooling. Moreover, the filling and emptying of the cooling chamber introduce another operation and expenses into the marketing chain (Rai and Arumuganathan 2008).

Ice bank cooling

In order to reduce the loss of mushroom weight during vacuum cooling, ice bank cooling of mushrooms is now in vogue wherein a stack of mushrooms is passed through the forced draft of chilled and humidified air coming from the water body maintained at sub zero temperature, called 'ice bank' (Rai and Arumuganathan 2008).

Radiation preservation

Radiation preservation offers a process of cold sterilization where the mushrooms are preserved with their natural characters. In this process, low doses of γ-irradiation are used to reduce microbial contamination and extend shelf-life of mushrooms. The γ-irradiation is given immediately after harvest for optimum benefits. The irradiation delays the after-harvest maturation of mushroom and reduces the loss of weight, colour, texture and nutritional quality. However, it has one drawback that the flavour decreases in irradiated mushroom (Staden 1967, Campbell et al. 1968, Wahid and Kovacs 1980, Roy and Bahl 1984, Lescane 1994, Roy et al. 2000). Cobalt 60 (Co_{60}) is used as a common source of γ-rays. A dose of 10 kGy (Kilo Gray) is effective to destroy microorganisms completely. The shelf-life of button mushrooms (*Agaricus bisporus*) is increased by a period of 10 days with the application of gamma rays of 2 kGy and storage at 10ºC (Lescane 1994). It is also observed that amino acids in fresh mushrooms are preserved better at low levels of γ-irradiation (Benoit et al. 2000). Thus, in keeping the quality of button mushrooms, a low dose of 1 kGy of electron-beam γ-irradiation is suggested for enhancing shelf-life of mushroom slices (Koorapati et al. 2004).

Transport of Fresh Mushrooms

Fresh mushrooms are transported to markets in cold condition. In order to keep the mushrooms cool during transport, the polypacks of mushrooms are stacked in small wooden cases or boxes with sufficient crushed ice in polypacks and that is over wrapped with paper. However, for large quantities of mushrooms transporting to long distance, refrigerated trucks are suitable, although, it is costlier than the former one.

Steeping Preservation

This method is simple and economical as compared to canning and freeze-drying. In this method, the mushrooms are preserved for a short period by steeping them in solutions of salt or acids. Traditionally, the mushrooms after harvest are washed in water or chemical added water. Thereafter, they are blanched for 5 minutes in brine solution and filled in cans with brine solution. However, the salt concentration more than 10 per cent in steeping solution intensifies browning reaction of the mushroom. This adverse reaction is nullified by adding organic acids (Wang and Cheng 1978). Thus, several improvements have been made by different workers during their times. Pruthi et al. (1984) have reported that dipping of mushrooms in dilute solutions of hydrogen peroxide for half an hour and then steeping in the solution containing 0.25 per cent citric acid and 550 ppm sulphur dioxide have significant effect on the whiteness of mushrooms. In other case, the steeping solution, consisting of 2 per cent sodium chloride, 2 per cent citric acid, 2 per cent sodium bicarbonate and 0.15 per cent Potassium Metabisulphite (KMS) is suggested to preserve blanched button mushrooms for 8–10 days at 21–28°C (Kapoor 1989). Alternatively, the steeping solution consisting of 2 per cent salt, 2 per cent sugar, 0.3 per cent citric acid, 0.1 per cent KMS and 1 per cent ascorbic acid is used for the same purpose (Singh et al. 1995). The fresh oyster mushrooms are preserved in steeping solution containing 5 per cent salt (sodium chloride), 0.2 per cent citric acid and 0.1 per cent KMS in glass bottles (Adsule et al. 1981). In this method mushrooms can be kept fresh for about 8–10 days. Milky mushroom can be stored for more than 180 days in steeping preservation (Sohliya et al. 2010). In that case, the fresh mushrooms are blanched in sodium hypochloride (NaOCl) of 100 ppm for 5 minutes before steeping in sodium chloride (common salt 6 per cent), ascorbic acid (1 per cent) and potassium metabisulphite (0.1 per cent) mixed solution.

The National Research Centre for Mushrooms (now Directorate of Mushroom Research) of Solan has suggested following two methods of steeping preservation (Annual Report, 2000–01, NRCM, Solan):

Unexhausted steeping preservation

In this method, fresh mushrooms are washed and blanched in 0.05 per cent KMS for 5 minutes. Now, the solution is drained out and the mushrooms are washed in cold water for 4–5 times. The mushrooms are then filled in bottles or cans. Hot brine of 18–20 per cent NaCl and 0.1 per cent citric acid is poured in the mushroom filled bottles or cans. The bottles or cans are closed with lids tightly and stored at room temperature. This type steeping can preserve mushroom up to 3 months.

Exhausted steeping preservation

In this case, the fresh mushrooms are washed, blanched and filled in bottles or cans with hot acid mixed brine solution similarly as practiced in the preceding method. However, here, instead of closing immediately with lids, the filled up bottles or cans are kept in hot water bath to raise the brine temperature to 85°C for 10 minutes. Thereafter, the bottles or cans are closed tightly for storage.

LONG-TERM PRESERVATION

Long-term preservation helps to store mushrooms for longer period than the normal storage period. This is done mainly to get mushrooms during off-season and export. Generally, three different methods, like canning, drying and pickling, are followed in this system.

Canning

The process of sealing foodstuffs hermetically in containers and sterilizing them by heat for long-term storage is known as **canning**. The canning is required to preserve mushrooms for period of a year or more and to export these in foreign countries, as most of the international trades in mushrooms prevailed in this form. The canning is done soon after harvest. In case, it is delayed then the mushrooms are stored at 4–5°C till processed. The button mushrooms keeping 1 cm stem after cutting or sliced longitudinally and paddy straw mushrooms of button or egg staged are generally canned (Mudahar and Brains 1982, Pruthi et al. 1984, Beelman and Edwards 1989). The canning process is completed by different units of operations, like cleaning, blanching, filling, sterilization, labelling, packaging and storage, and it requires certain infrastructure facilities.

Facilities required

Lid-embossing machine: This machine is required to emboss the lids with the reference letters or figures like date of expiry, rate, batch number, brand name and quantity. It is operated by foot treadle and embosses the lid by marker dies, without piercing the lid.

Can reformer: The machine is required to make the flattened can bodies to round, prior to flanging. During operation, the flattened can is mounted on the rubber roller. On lowering the padal, it presses the can against the rotating steel roller and thereby giving the can round in shape.

Can flanger: This is a hand operated machine used for simultaneous flanging of both sides of round can.

Flange rectifier: This is used to rectify the misshaped flanges of cans.

Steam jacketed fettle: This is used for batch heating and blanching of mushrooms in brine solution.

Exhaust box: This machine is consisted with a conveyor chain to keep the filled up cans in contact with steam for 1 to 2 minutes.

Double seamier: This machine is used for seaming the possessed cans as well as flanged cans with embossed lids on both sides.

Canning retort: This equipment is used for sterilization of cans under pressure. It is equipped with pressure gauge and safety valve.

Boiler with fittings and accessories: This is required for generating steam in exhaust box.

Hand refract metre: 1

Seam checking gauge: 1

Can tester: 1

Aluminum trays: Required for handling cans

Vacuum tester: 1

Water supply and drainage: This is required for cleaning mushrooms, brine preparation, washing floors and machinery, etc.

Method

Cleaning: The mushrooms with tight closed veil (for button mushroom) or volva (as in case of paddy straw mushroom) are selected for canning. These are washed 3–4 times in cold running water to remove adhering substances. Then, these are immersed in citric acid (0.1 per cent) or sodium metabisulphite (0.3 per cent) solution to prevent discolouration and taken out for blanching.

Blanching: Blanching removes gases from mushroom tissue and reduces microorganisms. It is done by cooking the mushrooms at 95°–100°C for 5–6 minutes in water containing 0.1 per cent citric acid and 1.0 per cent common salt in steel kettles (Tanga 1974, Saxena and Rai 1990). Six minute blanching is mostly used to give proper drained weight of mushrooms. Immediately after blanching, a short spray of cold water is followed to cool the mushrooms to or below 36°C.

Filling: After blanching the mushrooms are filled in tin cans. Different sizes of cans are used in our country to preserve mushrooms. For example, both A-1 tall can filled with 220 g mushrooms and A-2$^1/_2$ can filled with 440 g mushrooms are prepared for domestic marketing, while, A-10 can filled with 1960 g mushrooms is used for export. However, the standard international sizes of cans are considered as 16 oz of can for 454 g or 68 oz for 1930 g mushrooms, which are used by several mushroom growing countries. The cans are filled with brine solution of 2 per cent salt and 0.1 per cent citric acid or 100 ppm ascorbic acid, keeping 1.5 cm of head space. The temperature of the brine solution is maintained to 90°C during pouring. The brine solution with 2 per cent salt, 1 per cent sugar and 0.05 per cent citric acid may also be used for filling to obtain better result as suggested by Azad et al. (1987).

Lidding or clinching: The cans, after filling with mushrooms and hot brine solution, are covered loosely with the lids or clenched the lids partially to the cans by a single first roller action of a double seamier machine.

Exhausting and sealing: The exhausting is done to remove air from the can. The loosely covered cans are exhausted in exhaust box by keeping them on a conveyor belt which moves slowly in exhaust box. The cans are passed through exhaust box for 10–15 minutes in hot steam which comes from a boiler. The length of exhaust box is adjusted to couple with the speed of conveyor belt. When, the hot steam in exhaust box brings the brine temperature to 85°C at the centre of the can then that is sealed hermetically with double seamier. Alternatively, the cans are exhausted by keeping them in boiling water till the temperature in the centre reaches to 85–90°C. Then the cans are sealed hermetically with double seamier. After sealing, the cans are kept upside down.

Sterilization: Sterilization is done to prevent the spoilage of mushrooms by microorganisms. Generally two different methods are followed.

Steriflame sterilization—This is a continuous process in which the cans are passed over the gas burners for at least 3–8 minutes.

Autoclave sterilization—This is a batch process in which the cans are placed in a canning retort (autoclave) and sterilized at 10 p.s.i. for 25–35 minutes or at 15 p.s.i. for 15 minutes.

Cooling: The cans are cooled in cold water tank immediately after sterilization. This process helps to stop over cooking of mushrooms and to prevent stack burning. Further, the abrupt cold shock also kills the microorganisms, if any survived.

Labelling and packing: The cans after cooling are allowed to dry their outer surfaces. Then these are labelled manually or mechanically and packed in wooden crates or corrugated card board cartoons. The packed materials are stored in a cool and dry place.

Drying

Fresh mushroom contains about 90 per cent moisture. The high moisture in mushroom makes it perishable and creates favourable conditions for the growth of microbes and insects during storage. However, if the moisture level in mushroom is brought down to or below 10 per cent level, this can be stored for a longer period. After storage, the dried mushroom regains to a large extent its flavour and texture on re-hydration. Further, the dehydrated mushroom is a very important ingredient in making soups and pasta (Tuley 1996, Gothandapani et al. 1997). Most of the mushrooms, like oyster, paddy straw, shiitake, black ear, silver ear, etc. are stored for long time in this process. However, for button mushroom the dry preservation is not encouraging due to its blackening and irreversible change of texture on drying. Mushrooms are dried by various ways, like sun drying, cabinet air drying, dehumidified air-cabinet drying, osmo-air drying, freeze-drying, fluidized-bed drying and microwave drying.

Sun drying

Sun drying is the cheapest method of drying to preserve mushrooms for long time. In this method, the mushrooms are dried under direct sunlight. The harvested mushrooms, making into small pieces (if required), are dried in windy and bright sunlight on a metallic sheet. However, for better preservation, mushrooms are treated in the solution of 0.05 per cent KMS and 0.1 per cent citric acid mixture for about 10–15 minutes (Arunmuganathan et al. 2004). Alternatively, the mushrooms may also be pre-treated successively in 0.5 per cent KMS and in 0.5 per cent sodium benzoate solution for 15 minutes each in order to maintain nutritive contents well (Nehru et al 1995). Further, blanching in steam for 4–5 minutes prior dehydration is an eco-friendly measure of pre-treatment to preserve the sun dried mushroom better. In general, mushrooms are dried either by spreading them over trays or sheets in bright sun or beaded in a thin wire or thread and hanged in the air in sun. The mushrooms become dried within 12 to 14 hours in bright sunlight (Rama and John 2000).

Cabinet air drying

This type of drying is done in a machine, named cabinet air drier or tray drier. It consists of a heater with thermostat, a circulated air supply blower and a series of trays placed one above the other in a plenum chamber through which hot air at constant flow rate is allowed to pass. The fresh mushrooms are kept on trays in the machine and the doors are closed to run the machine. The drying temperature 55–60°C is mostly used to dry the mushrooms for 7–8 hrs (Rama and John 2000). This mechanically dried mushroom can be stored for more than a year (Bano et al. 1992). The colour of the machine dried mushroom is brown to creamy. However, some improvement in colour can be made by pre-treating the mushroom in the solution mixture of 0.1 per cent citric acid and 0.25 per cent KMS for 15 minutes (Arora et al. 2003). The dried mushrooms are packed in foil-laminated pouches for better storage (Kumar et al. 1980). The machine drying is very much effective during cloudy-rainy days. Further, the sun dried mushrooms are required to dry for 4–6 hrs in machine at 55–60°C before packing (Saxena and Rai 1990).

Dehumidified air cabinet drying

This drying technique is more or less similar to cabinet air drying, except that, here, moisture laden air of the cabinet is passed through a condenser (chiller) or desiccated medium, e.g. silica gel, to absorb moisture. Further, in this case the temperature 45°C is used which dried mushrooms faster than the preceding method, thus, giving better quality of mushrooms.

Osmo air drying

This is a novel technique, in which the mushrooms are kept at first in a concentrated salt solution, called osmotic syrup and latter, they are taken out and air dried. The common salt (NaCl) of 15 per cent is used to remove about 35 per cent initial moisture of button mushroom by osmosis in an hour (Kar and Gupta 2001). In case of milky mushroom, 25 per cent salt solution is used for 6 hrs to remove 25 per cent moisture (Amuthan et al. 1999). The osmosed samples required about 3 hrs to dry in air cabinet for producing final product. The time period required for drying depends upon the concentration of the osmotic solution used and the duration of osmosis. The colour of osmo air dried samples show less browning.

Freeze drying

This method is used in limited cases due to high cost involvement in drying. The major advantage in this method is that the original shape of mushrooms is retained with superior flavour. However, the dried mushrooms become very brittle and needs to pack in sturdy packings and cushion-packs flushed with nitrogen for keeping good quality. The main principle of freeze drying is that the water from mushrooms is removed at first by forming ice crystals and sublimation, and later, the remaining water is removed by evaporation. In practice, the mushrooms are made into slices at first. The slices are then immersed in a solution of 0.05 per cent KMS and 2 per cent NaCl for about 30 minutes. These are now blanched in boiling water for about 2 minutes and cooled. The products are then frozen at –20 to –22°C and the moisture is removed by sublimation at very low vacuum (0.012 mbar) for 12–16 hrs in a freeze drier machine (Kannaiyan and Ramaswamy 1980, Kapoor 1989). The freeze dried mushrooms contain 3 per cent moisture and are packed under vacuum.

Pickle Preparation

Preservation of mushroom in pickle preparation is very popular method in India. In this method mushrooms are preserved in acetified base of spices, salt and oil. Different compositions of ingredients are used to make tastes in pickle. Here, a few such compositions are described.

In conventional way, 500 g fresh mushrooms are put in a pan with sufficient amount (500 ml) of vinegar as such that the mushrooms are immersed in the acid. Thereafter, salt 28 g, mace 2 blades, ground ginger 4 g, chopped onion 15 g, pepper 2 g and nutmeg 1 g are added one after other and stirred. The whole preparation is now boiled for about 10 minutes and poured in a glass jar that finally closed by sealing. The vinegar in pickle is used for longer storage and taste of mushroom (Kannaiyan and Ramaswamy 1980).

Ingredients:

- Mushroom — 500 g
- Vinegar — 500 ml
- Salt — 28 g
- Mace — 2 blades
- Ginger — 4 g
- Onion — 15 g
- Pepper — 2 g
- Nutmeg — 1 g

Directorate of Mushroom Research (Solan) has advised different formulation for pickle preparation. In this case, the fresh mushrooms are washed at first in 0.05 per cent KMS solution, then sliced and blanched in 2 per cent salt and 0.1 per cent citric acid for 5–8 minutes. The blanched mushrooms are mixed with NaCl (10 per cent) for curing overnight. The excess water which oozed out is removed on the next day. The spices and preservatives are mixed to obtain desired taste and quality of pickle (Saxena and Rai 1990). The ingredients are as follows:

Ingredients:

- Mushroom — 1 kg
- Black mustard seed powder (rai) — 35 g
- Turmeric powder — 20 g
- Red chilly powder — 10 g
- Cumin seed powder — 1.5 g
- Fennel seed powder (saunf) — 1.5 g
- Aniseed powder (suwa/shopa) — 1.5 g
- Carom seed (ajwan) — 10 g
- Nigella seed (kalonji) — 10 g
- Mustard oil — 200 ml
- Salt — 90 g
- Acetic acid (as preservative) — As required
- Sodium benzoate (as preservative) — As required

The pickles as prepared may be kept for 1 year in a plugged glass or plastic jar.

In case of oyster mushroom good quality of pickle is prepared in a patented formulation by blanching the mushrooms in hot water at 80°C for 5 minutes (Arumuganathan et al. 2003, Rai and Arumuganathan 2008). The blanched mushrooms are rapidly cooled and mixed with 60 per cent brine to obtain the mushroom to brine ratio of 7:3 by volume. The mixture is then kept for 15 days at 15–20°C for fermentation. Then, the mixture is shifted to 0–4°C to stop fermentation and to get the pH 3.9 of the mixture. Now, sugar is added at the rate of 3.3 per cent of mushroom weight. Finally, the brine concentration reaches to 6.6 per cent in the product. Good quality of pickle is also prepared from paddy straw mushroom (Khader and Pandya 1981).

SOME INDIAN RECIPES OF MUSHROOM FOOD PREPARATION

Mushrooms just after harvesting may be used for food preparation (Figure 9.2). Here, some important recipes for preparation of food and preservation methods are given below:

Figure 9.2 Preparation of mushroom curry. **A** Cutting; **B** Adding spices and cooking; **C** Stirring and cooking; **D** Eating mushroom food with boiled rice.

Mushroom Soup

Mushroom soup is prepared by boiling 250 g mushrooms, a tomato and a little corn flour, all at a time, in water. When half of the water is evaporated then the cooked material is sieved through a clean gauge/cloth. A small piece of bread is fried in butter and immersed in the soup and this is stirred for mixing. The prepared soup is served as hot with a little butter.

In other way, the cut mushrooms (250 g) are boiled with grated ginger (15 g) and garlic (15 g) in three cups of water for about 10 minutes. These are cooled down and mixed in grinder/mixer. The mixture is now sieved through a cloth piece. Corn flour (20 g) is now added in milk (2 cup) and kept it separately. Now, butter (20 g) is placed on a heavy pan and heat it. There the sieved mushroom juice and the milk mixed corn flour are added and boiled for 10 minutes until it becomes thick. Required amount of sugar, salt and pepper (ground) are added to make taste (Verma and Rai 2005).

Instead, readymade mushroom soup powder (a popular value added product) may be used by boiling in equal amount of water. The soup powder is prepared by drying and grounding the mushroom into a powder. The mushroom powder (16 parts) is mixed with milk powder (50 parts) and corn flour (5 parts) and gelatinized in refined oil (4 parts). Now, the other ingredients, like salt (10 parts), cumin powder (2 parts), black pepper (2 parts), sugar (10 parts) and ajinomoto (2 parts) are mixed with the gelatinized mixture to prepare mushroom soup powder (Rai and Arunmuganathan 2008).

Mushroom Curry

Two big size tomatoes, onions (2 Nos.) and mushroom (200 g) are washed and cut into small pieces. Garlic (7 bulbs) and ginger (1 piece) are pasted. All these preparations are kept separately in containers. Now, cooking is done with onion and garlic in ghee/oil (1 large spoonful) until the materials turn to golden colour. Turmeric powder (1/2 tea spoonful), cumin powder (1/2 tea spoonful), chilli (as required for taste) and salt (as required for taste) are added to the cooking materials. Soon, the tomato pieces are added to the earlier materials and stirred until the tomato pieces melt and mixed well with other substances. One cup of curd or milk and the mushroom pieces are added there and these all are cooked for 5 minutes. During cooking the materials are stirred often, to mix these with each other. At last, water (if like) is added and boiled for about 5 minutes.

Mushroom Matar Masala

Mushroom (200 g) is cut into small pieces and kept separately. Then, green pea seeds are peeled out from pod (500 g) and kept separately. Onion (2 Nos.) and ginger (10 g) are cut into small pieces and kept separately. Now, the cooking is done with mustard oil/ghee or butter (2 tea spoonful) in a cooking pan. Onion, ginger, garlic (6 bulbs), cinnamon (small amount), chilli (as required for taste) and salt (as required for taste) are added in warm oil/ghee and stirred for about 5 minutes. Then the mushroom pieces and pea seeds (matar) are added and cooked for about 3–5 minutes. Water is added thereafter and boiled for 5 minutes. The cooked material is served hot along with some pieces of coriander leaves.

Mushroom Pakora

Mushroom (100 g) is washed and cut into small pieces. Slurry of besan (100 g, gram dal powder) is prepared with a little salt and cut pieces of green chilli according to taste. The mushroom pieces are mixed with the slurry and fried in oil/ghee. When the frying materials turned to light brown, pakoras are prepared. These are served with tomato sauce.

In other way, fresh mushrooms (500 g) are taken and boiled them in 1.5 litres of water with half table spoon full (tbs) salt for 5 minutes. The water is then drained out and mushrooms are squeezed to remove water. In a separate container the other ingredients, like chopped onion (1 big size), ginger paste (2 tbs), anar dana powder (1tbs), gram flour (150 g), salt (required amount), chopped green chillies (as required), garam masala 10 gm are mixed with a little water to make a thick paste. Now, the mushrooms are dipped in the paste and fry in cooking oil in pan. These pakoras are served hot with pudina chatney (Verma and Rai 2005).

Mushroom Omelet

An egg is broken and the inner material is mixed with salt and chilli pieces thoroughly and then that is kept separately. Two mushrooms and an onion are cut into small pieces, and are fried on a frying pan with oil/ghee or butter until they turned to brown. The fried materials are taken out and mixed with the egg slurry as prepared earlier by stirring. Oil or ghee is poured on a frying pan and the mixed materials are fried as omelet. This is served hot with tomato sauce.

Mui Barak (Tripuri Food)

Mushrooms (200 g) are cleaned and cut into small pieces. These are transferred to a cooking pan. Then, cut pieces of one big size onion, six to seven cut green chillies, four pieces of dry fish (sidal), a tea spoonful of salt and a half tea spoonful of turmeric powder are added to the mushroom in the cooking pan. Later, water is added and materials are boiled for 15–20 minutes. The cooked material is ready to serve with rice.

Mushroom Fish Curry

Cleaned cut pieces of mushrooms (200 g), potato (2 medium sized), onion (2 medium sized), chilli (6–7 nos.), tomato (1 No.), ginger small pieces, garlic 6–7 bulbs, fish 250 g (rohu or katla, peeled without scales and cut into small pieces) are required to prepare this food. At first, the fish pieces are fried by mixing a small amount of turmeric powder and salt and taken out to keep in a separate bowl. Now, in a cooking pan apply required amount of mustard oil, give onion pieces at first to fry for a while, and then add about half tea spoon full of fennel, fenugreek and black cumin mixture and stir. Add mushroom pieces, potato pieces, tomato pieces, chilli pieces, ginger paste and garlic paste, one after other or at a time as convenient and stir for mixing. Add cumin (1/3 spoon full), turmeric powder (1/2 spoon full), sugar (1/2 spoon full) and salt (as preferred). All stir well for mixing and cook up to the stage when all liquid matter is dried up and the cooking materials are approaching to stick with the cooking pan surface. Now, add water and fish fries to the cooking mixtures and boil together for about 5 minutes. The mushroom fish curry is ready as food with boiled rice.

CHAPTER

10 Economics in Mushroom Cultivation

Understanding of economics of mushroom cultivation is a very essential aspect to establish mushroom farm (small, medium or large) or to cultivate in commercial scale at farmers level. However, the calculation in determining uniform economics of mushroom cultivation is a difficult aspect to bring into accuracy. Since, the mode of mushroom cultivation varies from country to country, region to region and farm to farm. It is a highly technical and skilled activity, thus the production of mushroom in a unit area depends on the aptitude of the persons involved and the facilities provided in cultivation. Apart from this, the cost for developing infrastructural facilities, raw materials and labourers are different from place to place. The selling prices of mushrooms are also fluctuating, ranging from ₹ 50 to ₹ 300 during different seasons at different places of India. Moreover, special care is needed for marketing and storage of mushrooms as these are very delicate and perishable goods. Thus, before starting this venture one should have thorough knowledge in this field and should survey the market for proper disposal of the produce.

In this chapter, the economics as calculated by Directorate of Mushroom Research, Solan (H.P.), for different types of mushroom cultivations are presented.

ECONOMICS OF BUTTON MUSHROOM CULTIVATION

The economics of button mushroom cultivation as calculated by Dr. B. Vijay, Principal Scientist, DMR, Solan, is presented herewith (Training manual, 2010, Agartala). He showed 30 to 38 Lakhs profit for small farm with production capacity 200–250 mt per annum, while, ₹ 1.18 Crores profit for medium scale farm with production capacity 500 mt per annum and ₹ 5.07 Crores for large farm with production capacity 3000 mt per annum. In addition, in the current chapter, lower rate of conversion efficiency in mushroom production is also considered to calculate the profits, assuming the occurrence of probable yield variation during different times. Further, this will increase confidence of the grower by knowing the assured least profit of this venture. The expenditure on a mushroom farm can be divided into fixed assets (non-recurring) and recurring expenditure. The expenditure on fixed assets include the cost for land, developing infrastructure and to set up plant and machineries, like boiler, blowers, compost handling equipments, air-conditioning equipments, shelves, etc. On the other hand, the recurring expenditure includes the cost of raw materials, spawn, casing soil, packaging and canning materials, energy cost, pesticides, labour charges and salary of the employees, etc.

Economics of Export Oriented White Button Mushroom Farm

Infrastructure facilities required

1. Environmentally controlled growing room, 60 units of size 75 ft × 27 ft × 13 ft with floor area 121,500 sq ft
2. Pasteurization tunnels, 9 units, of size 70 ft × 12 ft × 13 ft with floor area 14,560 sq ft
3. Composting yard, 1 unit of size 300 ft × 120 ft × 14 ft with floor area 136,000 sq ft
4. Area for pre-wetting of straw, 1 unit of size 150 ft × 100 ft with floor area 15,000 sq ft
5. Casing soil chambers, 4 units of size 35 ft × 12 ft × 10 ft with floor area 1680 sq ft
6. Cooling machine (A/C Compressor) room/Generator room/Electric cabin, 1 unit of size 75 ft × 27 ft × 13 ft with floor area 2025 sq ft
7. Boiler machine room, 1 unit of size 18 ft × 15 ft × 12 ft with floor area 270 sq ft
8. Bunker for blower service and steam in jet, 1 unit of size 12 ft × 9 ft × 9 ft with floor area 1080 sq ft
9. Spawning area 1 unit of size 120 ft × 20 ft × 13 ft with floor area 2400 sq ft
10. Spawn lab., 5 units of size 20 ft × 18 ft × 12 ft with floor area 1800 sq ft
11. Canning unit 1 unit of size 75 ft × 54 ft × 13 ft with floor area 4050 sq ft
12. Can store room, 1 unit of size 75 ft × 27 ft × 13 ft with floor area 2025 sq ft
13. Cold room, 1 unit of size 37 ft × 27 ft × 13 ft with floor area 1000 sq ft
14. Office, 6 units of size 15 ft × 12 ft × 12 ft with floor area 1080 sq ft
15. Stores/tool room/workers room/canteen, 4 units of size 18 ft × 15 ft × 13 ft with floor area 1080 sq ft
16. Corridor, 1 unit of size 864 ft × 25 ft × 13 ft with floor area 21,600 sq ft

Total floor area required: 220,150 sq ft, i.e. 21,000 sq metres.

In addition to above land, there would be required places for casing soil dump, road, paths, etc. Total approximate land required for the project is around **6 hectares**.

Initial investment

(i) Cost for site: Land (about 6 hectares) procurement and its development require ₹ 3,000,000.

(ii) Cost compost unit:

1. Construction cost of pre-wetting area @ ₹ 50/per sq ft is ₹ 750,000
2. Construction cost of composting yard @ ₹ 100 per sq ft is ₹ 3,600,000
3. Construction cost of pasteurization tunnels @ ₹ 300 per sq ft is ₹ 2,268,000
4. Construction cost of casing soil pasteurization rooms @ ₹ 300 per sq ft is ₹ 504,000
5. Construction cost of bunker for blower service and steam inlet room @ ₹ 200 per sq ft is ₹ 216,000
6. Construction cost of spawning area @ ₹ 200 per sq ft is ₹ 480,000
7. Construction cost of boiler room @ ₹ 200 per sq ft is ₹ 54,000

Total cost for composting unit is **₹ 7,872,000**

(iii) Cost for production and canning facility:

1. Construction cost of cropping rooms @ ₹ 300 per sq ft is ₹ 36,450,000
2. Construction cost of canning hall @ ₹ 250 per sq ft is ₹ 810,000
3. Construction cost of A/C room/shed @ ₹ 200 per sq ft is ₹ 405,000
4. Construction cost of cold room @ ₹ 400 per sq ft is ₹ 400,000
5. Construction cost of can store room @ ₹ 200 per sq ft is ₹ 405,000
6. Construction cost of corridor in the cropping room @ ₹ 50 per sq ft is ₹ 1,080,000

Total cost for production and canning facility is ₹ 39,550,000

(iv) Construction cost of spawn laboratory: Laboratory with 5 rooms @ ₹ 250 per sq ft is ₹ 450,000.

(v) Miscellaneous cost: Construction of store/tool room/worker room canteen/office @ ₹ 200 per sq ft is ₹ 432,000.

GRAND TOTAL: (i) + (ii) + (iii) + (iv) + (v) = ₹ 5.13 Crores

(vi) Cost for plant and machinery:

Composting unit:

1. **Imported machineries = ₹ 1.00 crore**—(Compost turner, casing units, control switch panels, oxygen measurement equipment, central computer, special low voltage cable, tunnel pulling nets, tunnel gliding nets, net cleaning machines, floor grill moulds, ammonia measuring equipments, high pressure sprayers, filling conveyer plus feed hopper, oscillating head filling machine, tunnel emptying winch with combination of spawn dispenser and bagging machine, air handling units for the tunnels)
2. **Indigenous machineries = ₹ 40.00 lakhs**—(Boiler, tractor, trailer, doors of the tunnels and casing, soil chamber, truck, RCC gratings in the tunnels, front end loaders, electrical system in the composting yard, water pumps piping and valves, tube well, electrical system for the tunnels, lighting in the yard and tunnel area, overhead water tanks, casing soil blowers ducts and gratings

   Total cost for composting unit = ₹ 1.40 crores

Growing rooms:

1. **Imported machineries = ₹ 1.5 crores**—(Control switch panels computer control, CO_2 control, central computer, low voltage control cables, temperature/humidity equipments, casing soil trailer, air handling units)
2. **Indigenous machineries = ₹ 2.0 crores**—(Water chilling plant, low pressure boiler, hot water pumps and piping, ducting for AHU including chilled water pumps and piping, racks for growing rooms, doors for growing rooms, electrical panels and cables)

   Total cost for plant and machinery of growing rooms is **₹ 3.5 crores**)

Spawn laboratory and canning unit = ₹ 1.5 crores: (Equipments for spawn production, like laminar airflow, BOD incubators, autoclave, hot air oven, microscopes, air conditioner, imported automatic canning machine and quality control equipments, etc.)

Miscellaneous equipments and expenses = ₹ 1.0 crores: (Expenses for set up transformer, diesel generator set, street light and cables, structural steel for support, erection of machinery and piping, painting, freight and insurance and other miscellaneous equipments)

Total cost for plant and machinery:

1. Compost unit ₹ 1.4 crores
2. Growing rooms ₹ 3.5 crores
3. Spawn laboratory and canning unit ₹ 1.5 crores
4. Miscellaneous ₹ 1.0 crore

Total = ₹ 7.4 crores

Total initial project cost (non-recurring expenditure):

1. Cost for site ₹ 30 lakhs
2. Cost for development of infrastructure ₹ 4.83 crores
3. Cost for plant and machinery ₹ 7.40 crores

Total = ₹ 12.53 crores

Interest and depreciation of the non-recurring expenditure

1. For site of cost ₹ 30.00 lakhs the interest @ 15 per cent is **₹ 4.50 lakhs**
2. For development of infrastructure of cost ₹ 4.83 crores depreciation @ 5 per cent is ₹ 24.15 lakhs and interest @ 15 per cent is ₹ 72.45 lakhs, total **₹ 96.60 lakhs**
3. For plant and machinery of cost ₹ 7.40 crores depreciation @ 10 per cent is ₹ 74 lakhs and interest @ 15 per cent is 1.11 crores, i.e. total **₹ 1.85 crores**

Total interest and depreciation = ₹ 2.86 crores

Recurring expenditure

(i) Expenditure on salaries and wages:

1. Salaries to the manager and permanent staff = ₹ 10 lakhs
2. Wages to the labourers (approximately 200 @ ₹ 60/day) = ₹ 43 lakhs

Total salaries and wages is ₹ 53 lakhs

(ii) Expenses on raw materials:

1. Wheat + Paddy straw ₹ 7,000,000.00
2. Water ₹ 500,000.00
3. Casing soil ₹ 900,000.00
4. Fuel ₹ 900,000.00
5. Energy ₹ 4,600,000.00
6. Spawn ₹ 1,200,000.00
7. Poultry manure ₹ 1,200,000.00
8. Urea ₹ 300,000.00
9. Gypsum ₹ 250,000.00
10. Disinfectants and pesticides ₹ 200,000.00
11. Selling expenses freight and insurance ₹ 3,000,000.00
12. Repair and maintenance ₹ 1,000,000.00

Total = ₹ 21,050,000.00 or ₹ 2.00 crores (approx.)

Cost of production

1. Interest and depreciation of the non-recurring expenditure = ₹ 2.86 crores
2. Salaries and wages = ₹ 0.53 crore
3. Cost of raw materials = ₹ 2.0 crores

Total cost of production is ₹ 5.39 crores

Total mushroom production

1. Fresh mushroom considering 100 per cent capacity utilization and 18 per cent conversion rate is 2700 tons.
2. Fresh mushroom considering 100 per cent capacity utilization and 20 per cent conversion rate is 3000 tons.

Profit calculation of canned mushroom considering 20 per cent conversion rate

1. Cost of A-10 can = ₹ 20.00
2. Cost of canning and brine solution = ₹ 12.00
3. Cost of mushrooms of 2 kg drained wt., i.e. approx. 3 kg fresh wt. = ₹ 54.00

Total cost of a can filled in mushroom = ₹ 86.00

4. Sale price of one case containing 6 A-10 size cans is US $ 18 (1 US $ = ₹ 46.00)
5. Total A-10 cans produced from 3000 tons mushrooms = 975,000 cans, i.e. 162,500 cases
6. Sale price of 162,500 cases @ ₹ 828 per case = **13.45 crores**
7. Total production cost of 162,500 cases = **8.38 crores**
8. Net profit at 20 per cent conversion is ₹ 5.07 crores

Profit calculation of canned mushroom considering 18 per cent conversion rate

1. Cost of A-10 Can = ₹ 20.00
2. Cost of canning and brine solution = ₹ 12.00
3. Cost of mushrooms (2 kg drained wt. approx., i.e. 3 kg fresh wt.) = ₹ 60.00

Total cost for filling an A-10 can with mushroom = ₹ 92.00

4. Cost of 1 kg of fresh mushroom in canned condition = ₹ 30.66 or ₹ 31.00
5. Sale price of one case containing 6 A-10 size cans @ $18 (1$ = ₹ 46) = ₹ 828.00
6. Sale price of 1 kg fresh mushroom in canned condition = ₹ 46
7. Profit of 1 kg of fresh mushroom in canned condition = ₹ 15.00
9. Net profit at 18 per cent conversion is ₹ 36,450,000, i.e. **3.64 crores**

Economics of Medium Scale Farm

Infrastructure facilities required

1. Growing room, 16 units of size 80 ft × 18 ft × 13 ft with floor area 123,040 sq ft
2. Pasteurization tunnels, 2 units of size 45 ft × 12 ft × 13 ft with floor area 11,080 sq ft
3. Composting yard, 1 unit of size 100 ft × 60 ft × 14 ft with floor area 16,000 sq ft
4. Area for pre-wetting of straw, 1 unit of size 60 ft × 40 ft with floor area 2400 sq ft
5. Casing soil chambers, 2 units of size 20 ft × 10 ft × 10 ft with floor area 400 sq ft

6. Cooling machine (A/C compressor) room/generator room/electric cabin, 1 unit of size 45 ft × 18 ft × 13 ft, with floor area 720 sq ft
7. Boiler machine room, 1 unit of size 15 ft × 15 ft × 13 ft with floor area 225 sq ft
8. Bunker for blower service and steam in jet, 1 unit of size 30 ft × 9 ft × 9 ft with floor area 270 sq ft
9. Spawning area 1 unit of size 34 ft × 12 ft × 13 ft with floor area 408 sq ft
10. Spawn lab, 3 units of size 18 ft × 15 ft × 12 ft with floor area 810 sq ft
11. Canning unit, 1 unit of size 65 ft × 18 ft × 13 ft with floor area 1170 sq ft
12. Can store room, 1 unit of size 40 ft × 18 ft × 13 ft with floor area 720 sq ft
13. Cold room, 1 unit of size 15 ft × 15 ft × 13 ft with floor area 225 sq ft
14. Office, 3 units of size 15 ft × 12 ft × 12 ft with floor area 408 sq ft
15. Stores/tool room/workers room/canteen, 1 unit of size 20 ft × 15 ft × 13 ft with floor area 300 sq ft
16. Corridor, 1 unit of size 162 ft × 12 ft × 13 ft with floor area 1944 sq ft

Total floor area required: 40,030 sq ft, i.e. 4000 sq mt approx.

In addition to above land would be required for casing soil dump, straw storage, poultry manure storage, etc.

Total approx. land required for the project 1 hectare

Initial project cost

(i) Cost of land (about 1 hectare) and its development: Require ₹ 1,000,000.
(ii) Expenditure on construction of spawn laboratory and office: @ ₹ 225 sq ft is ₹ 300,000.
(iii) Expenditure on civil works for growing room and composting unit: It is ₹ 7,800,000.
Total expenditure on purchasing land and carried out civil works is ₹ 9,100,000

(iv) Expenditure on plant and machinery:

1. Compost and casing unit = ₹ 1,000,000: (Includes compost turner, filling line, air handling units, under stack blowers, boiler, doors, gratings of the tunnel, electrical system and water system)
2. Production unit = ₹ 6,000,000: (Includes water chilling plant, air handling unit low pressure boiler, racks, vapour proof lightings, electrical panels, cables and doors of the growing rooms)
3. Canning unit spawn laboratory and other equipments = ₹ 2,000,000

Total expenditure on plant and machinery is ₹ 9,000,000

(v) Total initial cost of the project is = ₹ 18,100,000:

Recurring expenditure

1. Wages and salary = ₹ 850,000
2. Raw materials = ₹ 2,500,000
3. Energy and fuel = ₹ 1,150,000

Total recurring expenditure is = ₹ 4,500,000

Interest and depreciation of the expenditure

1. For site of cost ₹ 10.00 lakhs the interest @ 15 per cent is **₹ 1.50 lakhs**
2. For development of infrastructure of cost ₹ 81 lakhs depreciation @ 5 per cent and interest @ 15 per cent are **₹ 16.20 lakhs**
3. For plant and machinery of cost ₹ 90.00 lakhs depreciation @ 10 per cent and interest @ 15 per cent are **₹ 22.50 lakhs**
4. On working capital of ₹ 45.00 lakhs the interest @ 15 per cent is **₹ 6.75 lakhs**

Total interest and depreciation = ₹ 46.95 lakhs or ₹ 47.00 lakhs

Cost of production

(i) For raw materials is ₹ 2,500,000
(ii) For energy and fuel is ₹ 1,150,000
(iii) For wages and salary is ₹ 850,000
(iv) Interest and depreciation are ₹ 4,700,000

Total cost of production is ₹ 9,200,000

Total production at 100 per cent capacity utilization

(i) At 15 per cent conversion is 420 tons
(ii) At 18 per cent conversion is 500 tons
(iii) At 20 per cent conversion is 560 tons

Cost of production

(i) Cost of production/kg at 15 per cent conversion rate is ₹ 22.00
(ii) Cost of production/kg at 18 per cent conversion rate is ₹ 18.40
(iii) Cost of production/kg at 20 per cent conversion rate is ₹ 16.40

Profit for fresh mushroom selling

(i) Net profit at 15 per cent conversion rate and sale price @ ₹ 35,000/ton is 54.6 lakhs
(ii) Net profit at 18 per cent conversion rate and sale price @ ₹ 35,000/ton is 83.0 lakhs
(iii) Net profit at 20 per cent conversion rate and sale price @ ₹ 35,000/ton is 1.04 crores

Profit for canned mushroom selling

Alternatively complete production can be canned and sold in the market.

Considering 18 per cent conversion rate total production of mushroom is 500 tons.

Now, one ton fresh mushroom gives 2727 cans of No.1 tall tins or 500 tons of mushroom will give 1,363,500 cans.

Economics:

Cost of empty can @ ₹ 9 per can is ₹ 12,271,500
Cost of canning and labelling (@ ₹ 3 per can is ₹ 4,090,500
Cost of packing material (₹ 15 per cartoon, total 57,000 cartoons) is ₹ 860,000
Total cost of 500 tons of mushrooms is ₹ 9,200,000
Total cost of canning is **₹ 26,400,000**

Sale price of one can in the market is ₹ 28.00

Total sale price of 1,363,500 cans is ₹ 38,178,000

Net profit of canned mushrooms at 18 per cent conversion rate is ₹ 11,778,000, i.e. 1.18 crores (approx.)

Economics of Small Scale Farm

Infrastructure facilities required

1. Growing room, 8 units of size 60 ft × 18 ft × 13 ft with floor area 18,640 sq ft
2. Pasteurization tunnels, 1 unit of size 40 ft × 9 ft × 12 ft with floor area 1360 sq ft
3. Composting yard, 1 unit of size 60 ft × 40 ft × 14 ft with floor area 12,400 sq ft
4. Area for pre-wetting of straw, 1 unit of size 40 ft × 20 ft with floor area 800 sq ft
5. Casing soil chamber, 1 unit of size 12 ft × 10 ft × 10 ft with floor area 120 sq ft
6. Cooling machine (A/C compressor) room/Generator room/Electric cabin, 1 unit of size 30 ft × 18 ft × 13 ft, with floor area 540 sq ft
7. Boiler machine room, 1 unit of size 15 ft × 12 ft × 13 ft with floor area 180 sq ft
8. Bunker for blower service and steam in jet, 1 unit of size 12 ft × 9 ft × 8 ft with floor area 108 sq ft
9. Spawning area 1 unit of size 18 ft × 12 ft × 13 ft with floor area 216 sq ft
10. Spawn lab., 3 units of size 18 ft × 15 ft × 12 ft with floor area 810 sq ft
11. Canning unit 1 unit of size 60 ft × 18 ft × 13 ft with floor area 1080 sq ft
12. Can store room, 1 unit of size 30 ft × 18 ft × 13 ft with floor area 540 sq ft
13. Packing room, 1 unit of size 12 ft × 12 ft × 13 ft with floor area 144 sq ft
14. Office, 1 unit of size 30 ft × 18 ft × 13 ft with floor area 540 sq ft
15. Stores/tool room/workers room/canteen, 1 unit of size 15 ft × 12 ft × 13 ft with floor area 180 sq ft
16. Corridor, 1 unit of size 108 ft × 15 ft × 13 ft with floor area 1620 sq ft

Total floor area required: 18,278 sq ft, i.e. 1700 sq mt approx.

In addition to above, land would be required for casing soil dump, straw storage, poultry manure storage, etc.

Total land required for the project around is 1.5 acres (approx.).

Expenditure on fixed assets

1. Cost of land (about 1.5 acres) and its development require ₹ 500,000
2. Expenditure on spawn laboratory is ₹ 350,000
3. Expenditure on compost and casing unit is ₹ 700,000
4. Expenditure on production and canning units is ₹ 3,300,000
5. Expenditure on office block is ₹ 250,000

Total expenditure on fixed assets ₹ 5,100,000

Expenditure on plant and machinery

1. Compost and casing unit: ₹ 700,000
   (Includes compost turner, filling line, air handling units, under stack blowers, boiler, doors, gratings of the tunnel, electrical system and water system.)

2. Production unit: ₹ 3,000,000
   (Includes water chilling plant, air handling unit low pressure boiler, racks, vapour proof lightings, electrical panels, cables and doors of the growing rooms)
3. Equipments for canning, spawn laboratory and others ₹ 1,000,000

Total expenditure on plant and machinery is ₹ 4,700,000

Cost of the project

1. On land ₹ 500,000
2. On civil works ₹ 4,600,000
3. On machinery ₹ 4,700,000

Total cost of the project ₹ 9,800,000

Recurring expenditure

1. Wages and salary ₹ 600,000
2. Raw materials and energy and fuel ₹ 1,400,000

Total recurring expenditure is ₹ 2,000,000

Interest and depreciation of the expenditure

1. For site of cost ₹ 5.00 lakhs the interest @ 15 per cent is ₹ 75,000
2. For development of infrastructure of cost ₹ 46.00 lakhs depreciation @ 5 per cent and interest @ 15 per cent are ₹ 920,000
3. For plant and machinery of cost ₹ 47.00 lakhs depreciation @ 10 per cent and interest @ 15 per cent are ₹ 1,175,000

Total interest and depreciation are ₹ 2,170,000, i.e. ₹ 2,200,000

Cost of production

1. For raw materials, energy and fuel is ₹ 1,400,000
2. For wages and salary is ₹ 600,000
3. Interest and depreciation are ₹ 2,200,000

Total cost of production is ₹ 4,200,000

Total production at 100 per cent capacity utilization

At 18 per cent conversion is 180 tons
At 20 per cent conversion is 200 tons

Cost of production

Cost of production/kg at 18 per cent conversion rate is ₹ 23.30
Cost of production/kg at 20 per cent conversion rate is ₹ 21.00

Profit

Net profit at 18 per cent conversion rate and sale price @ ₹ 40 per kg is ₹ 3,006,000
Net profit at 20 per cent conversion rate and sale price @ ₹ 40 per kg is ₹ 3,800,000

ECONOMICS OF OYSTER MUSHROOM CULTIVATION

Cultivation of oyster mushroom is mostly done by the seasonal growers in an ordinary house without installing any environment controlling device. It grows on various agricultural residues with or without any fermentation. The methodology of its cultivation is very simple, but different technologies for substrate preparation and its pasteurization or disinfection are used. Further, the substrates which are used at different places are varied. The nature of substrates is based on its mode of availability at low price in a given place where the oyster mushroom is grown. In many cases, the substrates are of wasting in nature. Thus, there is no uniform method to cultivate the oyster mushroom throughout the country. Hence, uniform economics of oyster mushroom cultivation for all the places is hard to work out.

Here, the economics for small, medium and large farms of South India, farm of NABARD (India) specifications, farms in North West India and farms in NEH India as suggested by Dr. R.C. Upadhyay, Principal Scientist, Directorate of Mushroom Research, Chambaghat, Solan (H.P.), during a training in Agartala (Training Manual, October 2010) is presented below.

Farms in South India

The economics of small, medium and large oyster farms of Karnataka has been worked out by Mamatha (1998), discussed as follows:

Economics of a large farm

(a) Non-recurring expenditure:

(i) Cost for development of infrastructure facilities

Civil work

1. Cost of land is ₹ 39,761
2. Cost of building is ₹ 103,013

Total cost of civil works is ₹ 142,774

Equipments, machineries and fixtures

1. Cost of machineries and equipment is ₹ 15,457
2. Cost of office equipments is ₹ 1002

Total cost of equipments, machineries and fixtures is ₹ 16,459
Total cost for development of infrastructure facilities is ₹ 159,233

(ii) Interest and depreciation of the non-recurring expenditure

1. Depreciation on buildings @ 5 per cent is ₹ 5151
2. Depreciation on machineries @ 10 per cent is ₹ 1646
3. Interest on fixed capital @ 14.5 per cent is ₹ 23,089

Total interest and depreciation are ₹ 29,886

(b) Recurring expenditure:

(i) Variable cost of production of oyster mushroom

1. Cost of raw material, straw, spawn poly bags, pesticides, transportation charges and miscellaneous expenditures is ₹ 62,209

2. Cost of energy is ₹ 13,010
3. Labour wages are ₹ 35,436
4. Interest on working capital is ₹ 6916

Total recurring cost of production is ₹ 117,571

(c) Cost of production:

1. Depreciation and interest on capital investment are ₹ 29,886
2. Recurring expenditure on mushroom production is ₹ 117,571

Total cost of production is ₹ 147,457

(d) Production of mushroom:

1. Mushroom production (kg) in 10,732 poly bags per annum 7235 kg
2. Price of mushroom @ ₹ 33.48 per kg fresh mushroom is ₹ 242,302

(e) Profit calculations:

1. Income from mushroom on selling is ₹ 242,302
2. Total cost of production is ₹ 147,457
3. Profit is ₹ 94,845

Economics of a medium farm

(a) Non-recurring expenditure:

(i) Cost for development of infrastructure facilities

Civil work

1. Cost of land is ₹ 26,808
2. Cost of building is ₹ 42,328

Total cost of civil works is ₹ 69,136

Equipments, machineries and fixtures

1. Cost of machineries and equipment is ₹ 9912
2. Cost of office equipments is ₹ 793

Total cost of equipments, machineries and fixtures is ₹ 10,705
Total cost for development of infrastructure facilities is ₹ 79,841

(ii) Interest and depreciation of the non-recurring expenditure

1. Depreciation on buildings @ 5 per cent is ₹ 2116
2. Depreciation on machineries @ 10 per cent is ₹ 1071
3. Interest on fixed capital @ 14.5 per cent is ₹ 11,577

Total interest and depreciation are ₹ 14,764

(b) Recurring expenditure:

(i) Variable cost of production of oyster mushroom

1. Cost of raw material, straw, spawn poly bags, pesticides, transportation charges and miscellaneous expenditures is ₹ 31,758
2. Cost of energy is ₹ 3677

3. Labour wages are ₹ 18,479
4. Interest on working capital is ₹ 3370

Total recurring cost of production is ₹ 57,284

(c) Cost of production:

1. Depreciation and interest on capital investment are ₹ 14,764
2. Recurring expenditure on mushroom production is ₹ 57,284

Total cost of production is ₹ 72,048

(d) Production of mushroom:

1. Mushroom production (kg) in 5518 poly bags per annum is 3719 kg
2. Price of mushroom @ ₹ 31.14 per kg fresh mushroom ₹ **115,793**

(e) Profit calculations:

1. Income from mushroom on selling is ₹ 115,793
2. Total cost of production is ₹ 72,048
3. Profit is ₹ 43,745

Economics of a small farm

(a) Non-recurring expenditure:

(i) Cost for development of infrastructure facilities

Civil work

1. Cost of land is ₹ 9259
2. Cost of building is ₹ 26,360

Total cost of civil works is ₹ 35,619

Equipments, machineries and fixtures

1. Cost of machineries and equipment is ₹ 11,098
2. Cost of office equipments is ₹ 679

Total cost of equipments, machineries and fixtures is ₹ 11,777
Total cost for development of infrastructure facilities is ₹ 47,399

(ii) Interest and depreciation of the non-recurring expenditure

1. Depreciation on buildings @ 5 per cent is ₹ 1318
2. Depreciation on machineries @ 10 per cent is ₹ 1179
3. Interest on fixed capital @ 14.5 per cent is ₹ 6873

Total interest and depreciation are ₹ 9369

(b) Recurring expenditure:

(i) Variable cost of production of oyster mushrooms

1. Cost of raw material, straw, spawn poly bags, pesticides, transportation charges and miscellaneous expenditures is ₹ 13,820
2. Cost of energy is ₹ 2070

3. Labour wages are ₹ 8392
4. Interest on working capital is ₹ 1518

Total recurring cost of production is ₹ 25,800

(c) Cost of production:

1. Depreciation and interest on capital investment are ₹ 9369
2. Recurring expenditure on mushroom production is ₹ 25,800

Total cost of production is ₹ 35,169

(e) Production of mushroom:

1. Mushroom production (kg) in 2487 poly bags per annum is 1538 kg
2. Price of mushroom @ ₹ 34.08 per kg fresh mushroom is ₹ 52,405

(f) Profit calculations:

1. Income from mushroom on selling is ₹ 52,405
2. Total cost of production is ₹ 35,169
3. Profit is ₹ 17,236

Farm of NABARD (India) Specifications

In view of simple cultivation technology and moderate investment required, funding institutions in India are providing financial assistance to potential oyster mushroom growers. The National Bank for Agriculture and Rural Development (NABARD) has recently approved unit cost specifications for oyster mushroom cultivation with low-input technology, which is given hereunder (Upadhyay 2010).

Estimate for the unit size of 400 bags per crop and 2 crops per annum

(a) Capital investment:

1. Cost of construction of shed 15′ × 15′ × 6′ (225 sq ft) @ ₹ 30 sq ft is ₹ 6750
2. Racks of 12 Nos. are of ₹ 1200
3. Bamboos for racks (80 Nos.) of ₹ 400
4. Sprayer of number one is of ₹ 950
5. Boiler drum of number one is ₹ 300
6. Bucket of number one is ₹ 50
7. Miscellaneous expenditure (for knives, trays, rope, etc.) is ₹ 200

Total capital investment is ₹ 9850

(b) Operational cost (per bag):

1. Straw 1.2 kg (dry) @ ₹ 0.70 per kg is ₹ 0.84
2. Spawn 5 per cent of dry wt @ ₹ 0.02/gm is ₹ 1.20
3. Polythene bags of size 14 inch × 22 inch and weight 50 g @ ₹ 65 per kg is ₹ 0.35
4. Chemicals and *sutli* (rope) are ₹ 0.10
5. Contingencies (transportation, miscellaneous expenditure, etc.) are ₹ 0.20
6. Provision for bad wastage @ 10 per cent is ₹ 0.27

Total operational cost (per bag) is ₹ 2.96, i.e. Say ₹ 3.00

Operational cost for 2 cycles of 400 bags @ ₹ 3.00 per bag is ₹ 2400.00.
Unit cost including fixed cost and operational cost (a + b) is ₹ 12,250.00

(c) Production and Income:

1. Mushroom yield/bag is 500 g
2. Yield/crop (500 g × 400 bags) is 200 kg
3. Income from selling mushroom @ ₹ 25 kg per crop is ₹ 5000
4. Total income from 2 crops per year is ₹ 10,000

Note: *The implementing agencies should invariably ensure that the following stipulations are met before extending financial assistance to promote this activity.*

1. *The beneficiaries should have adequate experience/undergone training in mushroom cultivation imparted by the Directorate of horticulture.*
2. *The availability of high quality infection free spawn from recognized source is needed to be ensured.*
3. *The availability of technical guidance/supervision from a competent agency should be ensured.*
4. *The implementing agencies should ensure that proper marketing facilities will be provided to the beneficiaries.*
5. *The hygienic conditions are to be given top most priority in the mushroom unit to avoid infection and crop failure.*

Farms in North West India

The National Research Centre for Mushroom has worked out the economics for oyster mushroom cultivation in (i) poly houses and (ii) mud houses with wheat straw as substrate (Upadhyay 1990) as under:

Poly house (20′ × 15′ ×10′) cultivation

(a) Non-recurring expenditure—₹ 27,850
(b) Recurring expenditure per crop—₹ 3588
(c) Expected yield (70 per cent BE) per crop—350 kg fresh mushroom
(d) Gross income per crop @ ₹ 20 per kg—₹ 7000
(e) Net profit per crop—₹ 3412
(f) Annual profit from 6 crops—₹ 20,472

Mud house (60′ × 20′ × 12′) cultivation

(a) Non-recurring expenditure—₹ 13,000
(b) Recurring expenditure per crop—₹ 11,000
(c) Expected yield per crop (60 per cent BE)—1200 kg fresh mushroom
(d) Gross income per crop @ ₹ 20 per kg—₹ 24,000
(e) Net profit per crop—₹ 13,000
(f) Annual profit from 6 crops—₹ 78,000

Farms in NEH India

Paddy straw based cultivation by cube-culture method has been standardized by Verma and his co-workers. They have worked out the economics of a low-input farm made of bamboo mat and thatched roofing (8 m × 3.5 m × 3.5 m) with a capacity of 150 cubes (size: 50 cm × 25 cm × 18 cm) per crop and 6 crops in a year (Verma 1980, Verma 1985, Chandra et al. 1995) as follows:

Non-recurring expenditure

1. Cost of polythene sheet (150 sq m)—₹ 500
2. Cost of sprayer machine—₹ 1100
3. Cost of mushroom house (made of wooden poles, bamboo mat and thatch roofing (size 8 m × 3.5 m × 3.5 m)—₹ 5000
4. Drum for boiling water (200 litre)—₹ 700
5. Wooden frame with pressing board—₹ 200
6. Wire cage (2 Nos.) for immerging straw in boiled water—₹ 400

Total non-recurring expenditure—₹ 7900

Recurring expenditure per annum

1. Depreciation cost @ 20 per cent per annum during 5-year life of inputs—₹ 1580
2. Interest @ 12 per cent on principal amount—₹ 948
3. Labour cost @ 15 man days per crop for 6 crop of 90 man days @ ₹ 35 man days—₹ 3150
4. Cost of paddy straw (2 tons)—₹ 2500
5. Cost of spawn (mushroom seeds, @ ₹ 8/bag)—₹ 7200
6. Cost of pesticides and chemicals—₹ 1000

Total recurring expenditure per annum—₹ 16,378

Expected yield

(2 kg/cube, 150 cubes per crop, 6 crops) per annum—1800 kg fresh mushroom

Gross income per annum

@ ₹ 30 per kg—₹ 54,000
Expenditure per annum—₹ 16,378

Net profit per annum—₹ 37,622

Apart from this, the economics of oyster mushroom cultivation in Tripura which has been worked out by Biswas and Singh (2008b) is presented below. In this case, the economics is calculated based on the mushroom growing in bags and keeping on racks.

Non-recurring expenditure

1. Mushroom house—₹ 50,000
   (cemented floor, bamboo mat wall with plastering and thatched roof, size 22 ft × 13 ft in production room.)
2. Wooden racks—₹ 21,000
   (Nos. 3, 18′ × 3′, 4 tiers, with 2 ft gaps and two can accommodate 192 packets of 20″ × 16″ size in production room.)

3. Water tank (cemented)—₹ 5000
4. Sprayer—₹ 3000
5. Bucket, basket, polythene sheet—₹ 3000

Total of non-recurring expenditure—₹ 83,000

Recurring expenditure per annum

1. Depreciation cost @ 15 per cent per annum—₹ 12,450
2. Interest @ 8.5 per cent on capital—₹ 7055
3. Polythene bags 15 kg @ ₹ 120/kg—₹ 1800
4. Spawn packets @ ₹ 5 per packet of 1152 packets—₹ 5760
5. Labour charge @ 50 per man days of 20 man days/month for 6 months—₹ 6000
6. Paddy straw 1.5 tons—₹ 1500
7. Fungicides, insecticides, formalin, etc.—₹ 2500
8. Miscellaneous—₹ 500

Total of Recurring Expenditure—₹ 37,565

Income on oyster mushroom production

The income in oyster mushroom production is ₹ 64,512 (August–January, six crops, 192 poly bags/crop, @ 0.70 kg/bed, total production 806.4 kg, sale price at present rate ₹ 80 per kg)

Expenditure

The expenditure in oyster mushroom production is ₹ 37,565

Profit

The profit in oyster mushroom cultivation is ₹ 26,947.

ECONOMICS OF MILKY MUSHROOM CULTIVATION

The economics of milky mushrooms has been worked out by Dr. R.P. Tewari, Ex. Director, National Research Centre for Mushroom (Now, Directorate of Mushroom Research), Chambaghat, Solan (HP), India (Training Manual, October 2010). The mushroom grows in tropical climatic condition. It is mostly cultivated in commercial scale in south India. The economics as calculated for the farm producing 50 kg fresh mushroom per day is as follows:

Production cost per annum

1. Cost of paddy straw of 36,500 kg is ₹ 36,500
2. Cost of spawn of 4600 kg is ₹ 230,000
3. Cost of polythene bags (500 kg) for growing is ₹ 38,500
4. Cost of PP bags (500 kg) for packing is ₹ 38,500
5. Cost of casing material is ₹ 10,000
6. Cost of thread ball is ₹ 2000
7. Cost of formaldehyde of 50 litres is ₹ 3750
8. Cost of carbendazim (bavistin) of 2.5 kg is ₹ 2000

9. Cost of Melathion/Nuvan of 2 litres is ₹ 1600
10. Cost of bleaching powder of 50 kg is ₹ 1000
11. Labourer wages @ ₹ 2500/month of 3 numbers for 12 months are ₹ 90,000
12. Cost of water is ₹ 2400
13. Cost of electricity is ₹ 3400
14. Transportation charges are ₹ 10,000
15. Rent of building or shed is ₹ 12,000
16. Miscellaneous expenditure is ₹ 12,000

Total production cost per annum is ₹ 493,650; say ₹ 500,000

Fresh mushroom production considering 50 per cent conversion rate (kg/annum) and profit calculation

1. Fresh mushroom produced considering 50 per cent conversion rate is 18,250 kg
2. Used for self consumption is 180 kg
3. Wastage of mushroom is 180 kg
4. Net weight of marketable mushroom is 17,890 kg, say 17,900 kg
5. Cost of production/kg is ₹ 27.93 or ₹ 28
6. Sale price @ ₹ 45/kg is ₹ 805,500
7. Net annual profit is ₹ 305,500

Fresh mushroom production considering 100 per cent conversion rate (kg/annum) and profit calculation

1. Fresh mushroom produced considering 50 per cent conversion rate is 36,500 kg
2. Used for self consumption is 180 kg
3. Wastage of mushroom is 360 kg
4. Net weight of marketable mushroom is 35,960 kg, say 35,950 kg
5. Cost of production/kg is ₹ 13.90 or ₹ 14
6. Sale price @ ₹ 45/kg is ₹ 1,618,200
7. Net annual profit is ₹ 1,118,200

Further, Biswas and Singh (2008b) have worked out the economics of milky mushroom cultivation using the same infrastructure prepared for oyster mushroom cultivation in Tripura. This is done in order to show additional income by growing milky mushroom in 22 ft × 13 ft growing room once during unfavourable season of oyster mushroom as follows.

Non-recurring expenditure

Calculated with oyster mushroom = Nil

Recurring expenditure

1. Polythene bags (3 kg, @ ₹ 120/kg)—₹ 360
2. Paddy straw (200 kg)—₹ 200
3. Spawn packets (@ ₹ 5/packet, 192 packets)—₹ 960
4. Labour charge (25 man days/crop, 1 crop @ ₹ 50/man days)—₹ 1250

5. Fungicides, insecticides, formalin, etc.—₹ 500
6. Cow dung manure, sand—₹ 50

Total of recurring expenditure—₹ 3320

Income on milky mushroom production

₹ 8050 (milky mushroom, 1 crop, during April–May, 192 packets, 0.6 kg/bag, total production 115 kg, @ ₹ 70 per kg)

Total expenditure

The total expenditure in milky mushroom cultivation is ₹ 3320.

Profit

The profit in milky mushroom cultivation is ₹ 4730.

ECONOMICS OF PADDY STRAW MUSHROOM CULTIVATION

The economics of paddy straw mushroom has been worked out by Biswas and Singh (2008b) with the mushroom cultivated in the same infrastructure developed for oyster mushroom in Tripura. They prepared the economics based on three times cultivation of the mushroom during rainy season in a year. The method as followed is traditional with little modification in bed shape (cube shaped) and mode of bed preparation (with chemical disinfected crumpled straw without bundle preparation). The paddy straw mushroom requires only 10–12 days for developing fruit bodies in first flush, producing 70–80 per cent of its total yield. Thereafter, two or three more flushes generally come at 7–8 days interval, if the climatic condition favours its development. The cost for development of infrastructure is not considered here as the cultivation of paddy straw mushroom is advocated as an additional income to earn with oyster mushroom.

Non-recurring expenditure

Nil (included with oyster mushroom)

Recurring expenditure

1. Polythene sheet for covering beds—₹ 3000
2. Labour cost (6 man days/crop of 10–15 days span, 3 crops in 2 months, @ ₹ 50 per man day)—₹ 900
3. Paddy straw (400 kg for 120 beds of 1cu ft, 40 beds/crop)—₹ 400
4. Spawn (120 packets @ ₹ 5/packet)—₹ 600
5. Gram dal powder (12 kg @ 30/kg)—₹ 360
6. Fungicides, insecticides, formalin, etc.—₹ 1000
7. Nylon thread (2 kg)—₹ 160

Total recurring expenditure—₹ 6420

Income on paddy straw mushroom production

₹ 9600 (during June and July, @ 0.80 kg/bed of 120 beds, total production 96 kg, sale price ₹ 100 kg)

Total expenditure

The total expenditure of paddy straw mushroom cultivation is ₹ 6420.

Profit

The profit in paddy straw mushroom cultivation is ₹ 3180.

Apart from this, Verma (2002b) has worked out the economics of paddy straw mushroom cultivation while cultivated in traditional method by the seasonal growers in low cost hut. In this method, in addition to arhar/gram dal powder, single super phosphate is also used as additive. The economics are as follows:

Capital investment

1. Cost for construction of hut (floor size 30 ft × 20 ft) @ ₹ 50 per square is ₹ 30,000
2. Cost of sprayer machine (one number) is ₹ 1500
3. Cost of drum/bucket (one number) is ₹ 300
4. Miscellaneous expenditure is ₹ 200

Total capital investment is ₹ 32,000

Operational cost

1. Cost of straw 30 kg/bed (dry weight) @ ₹ 0.50/kg is ₹ 15
2. Cost of spawn @ 1 per cent of dry weight, i.e. 300 g/bed is ₹ 12
3. Cost of single super phosphate (250 g/bed) is ₹ 1
4. Cost of dal powder (150 g/bed) is ₹ 4
5. Cost of pesticides, plastic sheets, etc. is ₹ 4
6. Approximate labour cost per bed for 10 days is ₹ 10
7. Miscellaneous expenditure is ₹ 4
8. Operational cost/bed is ₹ 50
9. Number of beds in the above mentioned space is 36
10. Number of crops in six months cropping period is 8 crops
11. Operational cost of 8 crops (with 36 beds/crop) is ₹ 14,400
12. Interest of the capital investment @ ₹ 14.5 per cent per annum is ₹ 4640

Total operational cost is ₹ 19,040

Production and profit

1. Yield per bed is 2.50 kg
2. Total production from 8 crops (with 36 beds/crop) is 720 kg
3. Total income on selling mushroom @ ₹ 40 per kg is ₹ 28,800
4. Net profit in 6 months is ₹ 9760

In addition to this, the author has also described the economics of paddy straw mushroom cultivation if cultivated that following improved method in wooden cage. The economics are as follows:

Capital investment

1. Cost of hut with floor area 30 ft × 20 ft (600 sq ft) @ ₹ 50 per sq ft is ₹ 30,000
2. Cost of wooden cage (size: 1 m × 50 cm × 25 cm) of 100 numbers @ ₹ 100/cage is ₹ 10,000
3. Cost of sprayer (1 number) is ₹ 1500
4. Cost of drum/bucket (1 number) is ₹ 300
5. Miscellaneous expenditure is ₹ 200

Total capital investment is ₹ 42,000

Operational cost

1. Cost of straw 15 kg/bed (dry weight) @ ₹ 0.50/kg is ₹ 7
2. Cost of spawn @ 1 per cent of dry weight, i.e. 150 g/cage is ₹ 6
4. Cost of fuel for hot water @ ₹ 5 per cage is ₹ 5
5. Cost of plastic sheet @ 2 m per cage is ₹ 10
6. Labour cost per cage for 10 days is ₹ 20
7. Miscellaneous expenditure is ₹ 2
8. Operational cost/cage is ₹ 50
9. Number of cages in the hut is 100
10. Number of crops in six months cropping period is 8 crops
11. Operational cost of 8 crops (with 100 cages per crop) is ₹ 40,000
12. Interest of the capital investment @ ₹ 14.5 per cent per annum is ₹ 6090

Total operational cost is ₹ 46,090

Production and profit

1. Yield per cage is 2.50 kg
2. Total production from 8 crops (with 100 cages per crop) is 2000 kg
3. Total income on selling mushroom @ ₹ 40/kg is ₹ 80,000
4. Net profit in 6 months is ₹ 33,910

CHAPTER

11

Ancillary Information

SOURCES OF MACHINERY OR EQUIPMENTS

Machinery

1. **Christiaens Machines B.V. Horst**, Witveldweg 106, 5961 ND Horst, The Nederlands, Ph. +31(0)773999560, Fax: +31(0)773999561, Website: www.christiaensgroup.com
2. **Double T Equipment Ltd.,** 2 East Lake Way N.E., Airdrie, AB, T4A2J3, Canada, Ph. 403-948-5618, In U.S.A. Ph. 1-800-661-9195, Fax: 403-948-4780.
3. **Kasper Tideman**, Resident Representative of 'Agrisystems', Gold Fields Plaza, Office No. 24, 45-Sasson Road, Opposite Wadia Cottage, Pune-411001.
4. **Ranjan Kapoor,** D.V. Agro Industries Pvt. Ltd., Resident Representative of Dalsem Vaciap BV, Holland, B-325, New Friends Colony, New Delhi-110085.
5. **Rajiv Malhotra,** Resident Representative of Turati, Italia-63, Lajpat Nagar-II, New Delhi-110024.
6. **Thilot Holland B.V.,** Hoofdstraat 11-17, 5973 ND Lottum, The Netherlands, Ph. +31-77-463 1774, Fax: +31-77-463 2648, Email: thilot@thilot.com.

Boilers and Blowers

1. **Lal Sons & Company**, A-45, Mayapuri Industrial Area, Phase-I, Behind Syndicate Bank, New Delhi-110064, India, Ph. 91-11-28116223/8115121/5853441/8117483/5224343(R), Fax: 91-11-28117483.
2. **Indcon Boilers**, D-9/6, Okhla Industrial Area, Phase-I, New Delhi-110020, Ph. 011-26815336, Fax: 011-26815337, Website: www.indconboilers.com.
3. **Indvent Fans Pvt. Ltd**., 111 A, Dr. S.C. Banerjee Road, Kolkata-700100 (West Bengal), India, Ph. +91-33-23537640/23511602, Fax: +91-33-23537640/23591223.
4. **Laxmi Boilers,** 12 A, Chandivali Industrial Estate, Saki Vihar Road, Andheri East, Mumbai, Maharashtra (North)-400072, Ph. 022-49253971, 85241093.
5. **Thermax Ltd.,** 9, Community Centre, Basant Lok, New Delhi-110057, Ph. 011-46087200, Fax: 011-26145311, Website: www.thermaxindia.com.

6. **Urjex Industries,** S-26, Industrial Estate, Partapur, Meerut-250 102, Telefax: 0121-2440597, Website: www.urjexboilers.net.
7. **Regional Facility Centre**, Chambaghat, Solan-173213 (HP).

Cabinet Drier

1. **Windsons Scientific Works,** 10 W, Dr. Ram Manohar Lohia Marg, Delhi-110006, Ph. +(91)-11-27511841.

Canning Machinery and Cans

1. **Mather & Platt (India) Ltd.,** 805-806, Ansal Bhawan, 16, Kasturba Gandhi Marg, New Delhi-110001, Ph. 011-23766008, 23766009, 23766010.
2. **Mariental India Pvt. Ltd.,** 7/58, South Patel Nagar, New Delhi, 110008, Ph. 91-11-25840741/43, Fax: 91-11-25841200, Email: mariental@gmail.com.
3. **B. Sen Barry & Company,** Reg. Off. 65/11, New Rohtak Road, New Delhi-110005; Sales Office: 60/34, Ground Floor, New Rohtak Road, Karol Bagh, New Delhi 110005, Ph. 91-11-65702298, 65702299, 28720553, 28726429, Fax: 91-11-28720553, Email: info@bsenbarry.com, rohit@bsenbarry.com, senbarry@gmail.com.
4. **Bajaj Process Pack Maschinen Private Ltd.,** 7/27, Jai Lakshmi Industrial Estate, Site-IV, Sahibabad Industrial Area, Gaziabad-201001, UP, Ph. +91-120-2775119/4372848, Fax: +91-120-2775137, Email: sales@bajajmachines.com.
5. **Divecha Glass Industries**, 249, Bal Rajeshwar Road, 185, Marg Mulund (W), Mumbai-400080, Ph. +91-22-25612548, 25603360, Fax: +91-22-25671711.
6. **GST Kolkata (India Head Quarter),** Plot Y-7, Block EP, Sector V, Kolkata-700091, Ph. +91(33)23572278/79, Fax: +91(33) 40089897, Website: www.gstweb.com. www.gstwebservice.com.
7. **Techno Equipments**, 31, Parekh Street, Girgaon, Mumbai-400004.
8. **Rollatainers Ltd.,** 13/6, Mathura Road, Faridabad-121003, Ph. 0129-2271709, Fax: 0129-2275392, Website: www.rolapak.com.
9. **The Tin Plate Co. of India Ltd.,** 4, Bankshall Street, Kolkata-700001, Ph. 033-22435401, Fax: 033-22204170, Website: www.tatatinplate.com.
10. **Poysha Industrial Co. Ltd.,** Tiecicon House, Dr. E. Moses Road, Mumbai-400001, Ph. 022-4964378, Fax: 022-4964379.
11. **Spinco Biotech Pvt. Ltd.,** No. 4, Vaidyaram Street, T. Nagar, Chennai, Tamil Nadu-600017, India, Ph. 91-44-24340174, Fax: 91-44-24340761, Website: www.spincotech.com.

Cooling Machines/Refrigerators/Insulation

1. **Kriloskar Pneumatic Co. Ltd.,** Hadapsar Industrial Estate, Pune 411013, India, Ph. 91-20-26970133, 26870341, Fax: 91-20-26870297, 26870514, Email: acr-compressors@kpcl.net.
2. **S.M. Refrigeration,** 15/66, (Ground Floor), Old Rajinder Nagar, New Delhi-110060, Ph. 25753963, 25743196 (O) 9350204228, Fax: 011-25755574, Email: sm.refrigeration@gmail.com.

3. **Delta Systems,** 103-Boolani Est., Near Monginis, B/41 new Link Rd., Andheri (W), Mumbai-400053, Ph. 022-26736634/55705027, Fax: 022-26736606, Email: sales@deltabombay.com.
4. **Fedders Lloyd Corporation Ltd.,** 159, Okhla Phase-3, New Delhi-110020, Ph. 91-11-40627200, 40627300, Fax: 91-11-41609909, Email: info@feddersllloyd.com.
5. **Batliboi & Company Ltd.,** Bharat House, 5th Floor, Mumbai-400001, Maharashtra, Ph. 022-56378200.
6. **Premier Insulators**, 30 Westbrook Ave, Takanini, Auckland, New Zealand, P.O. Box 202046, Southgate, Takanini, Auckland.
7. **Thermo Systems Pvt. Ltd.,** Plot No. 21/C, IDA Phase-III, Jeedimetla, Hyderabad-500055, Ph. +91-40-44585800, Fax: +91-40-414585801.
8. **Airef Engineers Pvt. Ltd.,** 117, Tagore Park, Model Town, Delhi-110009, Ph. 011-27442508, Website: www.airef-engineers-pvt-ltd.
9. **Shah Engineering Agency,** F-6, Abhishek Complex, Opp. Amola Chambers, C.G. Road, Navrangpura, Ahmedabad-9, Ph. 079-26423444.
10. **Deepak Bhalla & Associates,** M-20, Lajpat Nagar-3, New Delhi-110024.

Cold Room

1. **Shakti Industries**, D-141, Industrial Focal Point, Patiala, Ph. 098148-15661, 09814711026, Email: industries.shakti@rediffmail.com.
2. **Indo Scientific & Surgicals,** 95A, Chittaranjan Avenue, Kolkata 700073, Ph. 22259253, Fax: 913322368655.

Laminar Air Flow/Autoclave/BOD/Oven/Other Instruments

1. **Grover Enterprises**, D-190, Sector-10, Noida-201301 (Delhi NCR), India Ph. +91-120-4224190 (5 Lines), Email: advertise@scientificbazaar.com.
2. **Indo Scientific & Surgicals**, 95A, Chittaranjan Avenue, Kolkata-700073, Ph. 22259253, Fax: 91-33-22368655.
3. **International Scientific and Surgicals**, Dev Building, The Mall Road, Solan-173213 (HP). Ph. 1792-220520.
4. **Klenzaids Engineers Pvt. Ltd.,** A-21, Midc Inds. Area, Marol, Andheri East, Mumbai-400093. Ph. +91-22-28323113, 28346036, Fax: +91-22-28375187.
5. **Narang Scientific Works Pvt. Ltd.,** G1-111, Ph-2, Mayapuri Industrial Area, Delhi-110064, Ph. +91-11-28112608, 28111589, 28112487, Fax: +91-11-28112608, Website: www.nsw_india.com.
6. **Prithvi Contamination Control Systems**, No. 13, Near Balaji Kalyana Mantapa, 5th Main, Banashankari 3rd Stage, Bangalore-560085, Ph. +91-9964036030, 9844084498.
7. **Saver Electronics & Instruments**, Reg. Office: 1442, Vazir Nagar, Kotla Mubarakpur, New Delhi-110003.
8. **Yorko Sales Pvt. Ltd.**, 11, Netaji Subhash Marg, Darya Ganj, New Delhi-110002. India, Ph. +91-11-23278306/23286624, Fax: +91-11-23264042.
9. **S.V. Instruments Analytica Pvt. Ltd.,** 3269, Ranjit Nagar, Near Pusa Gate, New Delhi-110008, Ph. 91-11-25843571, Fax: 91-11-25840571, Email: info@svianalytica.com, Website: www.svianalytica.com.

Microscope

1. **Olympus Corporation**, Shinjuku Monolith, 3-1, Nishi Shinjuku 2-chome, Shinjuku-ku, Tokyo, Japan, Website: www.olympus.com.

Glasswares/Plasticwares/Chemicals

1. **Nilkamal Crates & Containers,** Nilkamal House, 77/78, Marol Industrial Area, Road No. 13/14, MIDC, Andheri East, Bombay-400093, Website: www.nilkamal.com.
2. **Tarsons Products Pvt. Ltd.,** Jasmine Towers, Suite No. 213-214, 31-Shakespeare sarani, Kolkata, Ph.: 033-22892952-55, Fax: 033-22892956, Email: info@tarsons.in, Website: www.tarsons.in.
3. **GCC Biotech (India) Pvt. Ltd**., 351, BMK, Giri Nagar, Kalkaji, New Delhi-110019, Ph. 91-11-26440301, Fax: 91-11-26418606, Email: info@gccbiotech.com.
4. **Borosil Glass Works Ltd.,** Khanna Construction House, 44, Dr. R.G. Thadani Marg, Worli, Mumbai-400018, Ph. 022-4930362, Email: science@borosil.com, Website: www.borosil.com.
5. **Sisco Research Laboratories Pvt. Ltd.,** 26, Navketan Industrial Premises Co-op. Society Ltd., Shanti Nagar, Mahakali Caves Road, Andheri (E), Mumbai-400093, Ph. 91-22-42685800, Fax: 91-22-42685801, Email: info@srlchem.com, Website: www.srlchem.com.
6. **VWR Lab Products Pvt. Ltd.,** 6, Richmond Road, Bangalore-560025, India, Fax: 91-80-41237117, Email: vwr_india@vwr.com.
7. **Merck Specialities Pvt. Ltd.,** Shiv Sagar Estate 'A', Dr. Annie Besant Road, Worli, Mumbai-400018, Ph. 02266609137, Fax: 022-24954590, Email: Isa@merck.co.in, Website: www.merckspecialities.com.
8. **Technolab Scientific Co.,** 17/1D, Gopal Nagar Road, Kolkata-700027, Ph. 24799039, Fax: 24790626, Email: aspts@cal3.vsnl.net.in.
9. **Qualigens Fine Chemicals,** GlaxoSmithKline Pharmaceuticals Ltd., Dr. Annie Besant Road, Worli, Mumbai-400030, Ph. 24959595, Fax: 24962573, Email: gfc.x.customercare@gsk.com.
10. **Ablaze Glass Works Pvt. Ltd.,** E-52/2, Sardar Industrial Estate, Ajwa Road, Vadodara-390019, Gujarat, India; Ph. 91-2655593211, Fax: 91-2652564341, Email: ablazeglasslab@sify.com, Website: www.ablazeglassworks.com.
11. **Titan Biotech Ltd.,** A-2/3, 303305, Lusa Tower, Azadpur Commercial Complex, Azadpur, Delhi-110033; Ph. 91-11-27677960, Fax: 91-11-27674181, Email: akk_titanbiotechltd@yahoo.co.in, Website: www.titanbiotechltd.com.

MUSHROOM RESEARCH INSTITUTES, CENTRES AND LABORATORIES

1. **Brahma Spawn Production Centre**, Saheed Nagar, Bhubaneshwar, Orissa.
2. **Central Food Technological Research Institute,** Mysore-570013, Karnataka.
3. **Centre for Advanced Studies in Botany**, University of Madras, Guindy Campus, Madras-600025, Tamil Nadu.

4. **Centre of Tropical Mushroom Research & Training,** Department of Plant Pathology, O.U.A.T., Bhubaneshwar-751003.
5. **Department of Botany,** Calcutta University, 35-Ballygunge Circular Road, Kolkata-700019, West Bengal.
6. **Department of Botany,** Punjab University, Chandigarh-160014.
7. **Department of Microbiology,** Punjab Agricultural University, Ludhiana-141004, Punjab.
8. **Department of Mycology & Plant Pathology**, Banaras University, Varanasi-221005, Uttar Pradesh.
9. **Department of Plant Pathology**, Bidhan Chandra Krishi Vishwavidhalaya, Mohanpur, Nadia.
10. **Department of Plant Pathology,** College of Agriculture, Rajendranagar, Andhra Pradesh Agricultural University, Hyderabad.
11. **Department of Plant Pathology,** College of Agriculture, Mahatma Phule Krishi Vidyapeeth, Pune-411005.
12. **Department of Plant Pathology,** College of Agriculture, Kerala Agricultural University, Vellayani-695522.
13. **Department of Plant Pathology,** Tamil Nadu Agriculture University, Coimbatore-641003.
14. **Department of Plant Pathology,** Dr. Y.S. Parmar University of Horticulture and Forestry, Nauni, Solan-173230, Himachal Pradesh.
15. **Department of Plant Pathology,** Govind Ballabh Pant University of Agriculture and Technology, Pantnagar-263145, Uttar Pradesh.
16. **Department of Plant Pathology,** I.G.K.V.V., Raipur-492012, M.P.
17. **Department of Plant Pathology,** Uttar Banga Krishi Vishwavidhalaya, Pundibari, Coochbehar, W.B.
18. **Directorate of Extension Education,** CCS, Haryana Agricultural University, Hissar-125004.
19. **Directorate of Mushroom Research,** Chambaghat, Solan-173213, Himachal Pradesh.
20. **Division of Mycology & Plant Pathology,** Indian Agricultural Research Institute, Pusa, New Delhi-110012.
21. **Division of Plant Pathology,** Indian Institute of Horticultural Research, Hessaraghatta, Lake Post, Bangalore-560089, Karnataka.
22. **ICAR Research Complex for NEH Region**, Umroi Road, Umiam-793103, Meghalaya.
23. **ICAR Research Complex for NEH Region, Tripura Centre,** Lembucherra-799210, Tripura.
24. **Krishi Gyan Kendra,** Kothi No. 430, Sector-13 Urban Estate, Kurukshetra-132118.
25. **Maharashtra Association for Cultivation Sciences,** Erandwana, Pune-411004.
26. **Mushroom Development Centre**, P-Sector, PO: Itanagar, Directorate of Horticulture, Govt. of Arunachal Pradesh.
27. **Mushroom Development Project Canchipur**, Manipur-795003.
28. **Mushroom Laboratory,** Akhaura Road, Ramnagar-1 (by lane-3), Agartala, Tripura.
29. **Mushroom Laboratory,** Department of Plant Pathology, Rajasthan College of Agriculture, Maharana Pratap University of Agriculture and Technology, Udaipur 313001.

30. **Mushroom Laboratory,** Govt. of Meghalaya, Upper Shillong, Meghalaya.
31. **Mushroom Research Laboratory,** Department of Plant Pathology, College of Agriculture, Pune-411005, Maharashtra.
32. **Mushroom Research Laboratory,** Department of Plant Pathology, Tamil Nadu Agricultural University, Coimbatore-641003, Tamil Nadu.
33. **Mushroom Research Laboratory,** Gobind Ballab Pant University of Agriculture and Technology, Pantnagar-263145.
34. **Mushroom Research Laboratory**, Marchak, PO: Ranipul, East Sikkim.
35. **Regional Agriculture Research Station,** Kerala Agricultural University, Kumarakom, Kottyam-686566.
36. **Regional Research Laboratory,** CSIR, Srinagar-190005, Jammu and Kashmir.
37. **Spawn Laboratory,** Directorate of Horticulture, Kohima, Nagaland.
38. **Spawn Production Laboratory,** Lalmandi, Srinagar, Kashmir.
39. **Spawn Production Laboratory,** Talab Tillo, Jammu.
40. **Spawn Production Unit,** State Horticultural Department, Bhubaneshwar, Orissa.
41. **Teg's Muhrado Ltd.,** P.O.: Sadhupul, Cheonthe, Solan-173215, H.P.
42. **Vegetable Seed Production Centre**, Directorate of Horticulture, Govt. of Tripura, Nagicherra, Tripura.

MUSHROOM CULTURE INSTITUTES

1. **Directorate of Mushroom Research,** Chambaghat, Solan-173213, Himachal Pradesh.
2. **Division of Mycology & Plant Pathology,** Indian Agricultural Research Institute, Pusa, New Delhi-110012.

MUSHROOM FARMS

1. **Agro Dutch Industries Ltd.,** Farm: Village: Tofapur (near Lalru), Teh. Rajpura, Distt. Patiala, Punjab, Tel. +91-1762-505201, 505212, 505214, Telefax: +91-1762-505231, Email: enquiry@agrodutch.com; Reg. Office: S.C.O. 3D. 2nd Floor, Sector 33-D, Chandigarh 160020, India, Tel. +91-172-266336, 2606575, Telefax: +91-172-2604045, Email: enquiry@agrodutch.com.
2. **Aipine Solvex Group of Industries,** Indore, Madhya Pradesh.
3. **Cannaries,** 151, Rasulabad, P.O. Cavalary Lines, Allahabad (U.P.).
4. **Flx Food Ltd.,** M-32, Commercial Complex, Greater Kailash-II, New Delhi-110049.
5. **Ganga Singh Cold Stores,** Mushroom Farms, Bhogpur, Dist. Jullandhar, Punjab.
6. **Ginny Mushrooms,** A-36, Kailash Nagar, Sangrapura, Surat-3950002.
7. **Himra Mushrooms,** Hemraj Sheet Griha, Chakan, Dist: Pune, Maharashtra.
8. **Inkq Foods Pvt. Ltd.,** Ropar Road Nalagarh, Distt. Solan (H.P.), Ph. 01795-222907.
9. **Mahadeo Mushroom Farm,** Kishanganj-855107, Bihar.
10. **Mandeep Mushrooms (P) Ltd.,** 152, Sector-17, Gurgaon.
11. **Mohkan Mushroom Pvt. Ltd.,** Rajadas Vali Ram Building, Kirana Bazar, Akola-440001.
12. **Moon Mushroom Farm**, Rampur Industries, Rampur, Dist. Nainital (U.P.).

13. **Mousse Agro Farms Pvt. Ltd.,** Tigadi-591171, Karnataka.
14. **Pioneer Mushrooms,** 5-162, Bhavani Nagar, Moosa Pet, Hyderabad.
15. **Ponds India Ltd.,** (Mushroom Division), 26, C-in-C Road, PO Box No. 6809, Madras-600105.
16. **Royal Mushroom Farm,** Kadaktola, Rongpur, Silchar-788009, Assam.
17. **Ruby Mushroom & Canning,** Kurali, Dist: Ropar, Punjab.
18. **Sahas Agro Ltd.,** 5-6, Citicentre, Begum Bridge Road, Meerut-250002.
19. **Saptrishi Agro Industries Ltd.,** 15-A, First Avenue, Sastri Nagar, Adyar, Madras-600020.
20. **Tata Tea Ltd.,** Mushroom Division, P.O. Box 17, Munnar-685512, Kerala.
21. **Teg's Masrado Ltd.,** F-17, Hauz Khas Enclave, New Delhi-110016.
22. **Thakur Mushroom Farm,** Chambaghat, Solan-173213, H.P.
23. **Vikas Mushroom Farm,** Shamlaich, P.O. Barog Railway Station-173212, Distt. Solan (H.P.), Telefax: 01792-227651, Ph. 094180-27651, Email: vikasmushroom@gmail.com, Website: www.vikasmushrooms.com.
24. **Yo Mushroom Farm,** Chamakripul, Dhundan, Near MRF Plant, N.H. 88, Arki District, Solan, Himachal Pradesh-171102, Email: yomushroomfarm@gmail.com.

MUSHROOM MEDICINE PRODUCER

1. **Naturelife Healthcare Pvt. Ltd.,** Rajendra Tower-1, Asaltpur, Opp. Janak Puri Gurdwara, A-2 Block, WZ-13D/3, Ph. (O) 9810067307.

MUSHROOM EXPORTERS

1. **Aditi Exports,** G-13, Shreeji Palace, Navrang School Road, Navjivan, Ahmedabad-380014, Gujarat, India, Ph. +91-79-7424602/7424699, Fax: +91-79-7470505.
2. **Ambika Arihant Foods Pvt. Ltd.,** No. 187, H.S.I.D.C. Kundli, Near Post Office, Phase No. 2, Sonepat, Haryana-131028 (India), Ph. +91-11-20338171, +91-9811043219.
3. **Alto Industrial Corporation,** 2788, Gali Rajputana Subzi Mandi, Delhi-110007.
4. **Ankur International,** 2147, Sector 35C, Chandigarh-160036.
5. **Asian Agro Pvt. Ltd.,** B-3, Part-1, Gujranwala Town, New Delhi-110009. Ph. 2715774 (4 Lines), Website: www.waterfry.com/asian-agro-pvt-ltd.
6. **Asiatic Agencies,** 4850, 24, Ansari Road, Darya Ganj, New Delhi-110002, Ph. +91-11-23275670, +91-11-23274337.
7. **Bombay Grain Crushing & Spice Mills,** 502, Vardhman Chambers, 5th Floor, 127C, Kalyan Street, Danabunder, Bombay-400009.
8. **Calcutta Cellulose Pvt. Ltd.,** 4C, Meherali Mondal Street, Kolkata-700027.
9. **Chadha General Store,** SCF No. 3, Sector 16D, Chandigrah-160016. Ph. 0172-2770546.
10. **Dhanuka Exports Pvt. Ltd.,** M-279, Greater Kailash, Part-II, New Delhi-110048.
11. **Double Dee,** P-744, Block-A, Lake Town, Kolkata-700089. Ph. +91-33-25219285, Fax: +91-33-2521 4107, Email: doubledee@duttadasgupta.com.
12. **Elmac Foods,** Elmac House, 20/1B, A.K. Md. Siddiqui Lane, Kolkata-700016, India.

13. **Emcees Commercial Company,** P.O. Box 3162, 117/119, Kazi Sayed St., Mumbai-400023, Ph. +91-33-22444778/22460183, Website: http://www.elmacfoods.com.
14. **Flex Foods Ltd.,** M-32, Commercial Complex, Greater Kailash-II, New Delhi-110048 and A-108, Sector-4, Noida-201301, Uttar Pradesh, Ph. +91-120-4012345/2459444, Fax: +91-120-2442909, Email: info@flexfoodsltd.com.
15. **Flour & Foods Ltd.,** 10/11, Yeshwant Niwas Road, Indore-452001, India, Ph. 91-731-537365/537639, Fax: 91-731-430370.
16. **Ganga Kaveri Seeds Pvt. Ltd.,** Reg. Office: BF/5-D, DDA Flats, Munirka, New Delhi-110067, Address: Suit 1406-1407, Babu Khan Estate, Bashir Bagh, Hyderabad, Andhra Pradesh-500001, India, Ph. 040-242450, Fax: 040-233418.
17. **Godrej Agrovet,** Pirojshanagar, Eastern Express Highway, Vikhroli (E) Mumbai-400079, India, Ph. +91-22-25188010, Fax: +91-22-25188485, Website: www.godrejagrovet.com.
18. **Haryana Agro Food Processing Plant,** G.T. Road, Murthal, Sonepat-131001, Haryana. Ph. 2053 2092.
19. **Jayaex,** 8, Nattupillaiar Kovil, Street, Chennai-600001.
20. **Kumkum Food Products Pvt. Ltd.,** 278, Bhat Bazar, Narshi Natha Street, Mumbai-400009, Maharashtra, Ph. +91-22-28558577/28557855.
21. **Marex Corporation,** 22, 4th Floor, Moti Lane, Kolkata-700013.
22. **R.K. Proteins & Foods Pvt. Ltd.,** B-164, Industrial Area, Phase-1, Okhla, New Delhi-110020, Ph. +91-11-26816397.
23. **Saraf Foods Ltd.,** 720-721 GIDC Estate, Vaghodia-391760 Dist. Vadodara. India, Ph. +91-2668-262140, Fax: +91-265-2337397, Email. info@saraffoods.com.
24. **Sharma Brothers,** 20, Amba Market, Ambala-134003, Haryana, Ph. +91-171-2440082.
25. **Simran Export Inc.,** A-51, Ph. 2, Mangolpuri Industrial Area, Near Peera Garhi Chowk, New Delhi-110034, India, Ph. +91-11-47051801, +91-11-47051802, +91-11-47051803, +91-9811016875, Fax: +91-11-47511346, +91-11-25224087, Email. simran@simranindia.com, simran@del2.vsnl.net.in, Website: www.simranidia.com.
26. **Sonkem India Pvt. Ltd.,** 115, Prabhat Kiran, 17, Rajendra Place, New Delhi-110008, Ph. 91-11-25763092/25716172, +919810001032, Fax: 91-11-25755269.
27. **Usha International India,** 8, Commercial Centre, Malcha Marg, Diplomatic Enclave, New Delhi-110021.
28. **Zuari Foods & Farms Pvt. Ltd.,** 373, Opp Megsons Super Centre, Dayanand Bondodkar Marg, Miramar, Goa 403001, Ph. +91-832-2465222, 2465111.

Glossary

Adnexed: Gills attached slightly with stipe.
Agaric: Gill fungus.
Amyloid: Greyish, purple, bluish or blackish violet when treated with Melzer's reagent containing iodine, potassium iodide and chloral hydrate.
Anastomosis: Undergoing hyphal fusions with one another.
Annulate: With annulus or ring.
Annulus: A ring-like portion of the ruptured marginal veil, after the expansion of pileus.
Apiculus: A part at the end of spore where it is attached to the sterigmata.
Applanate: Pileus is almost flat.
Areolate: Surfaces of cap and stem showing cracks or crevices in some places.
Ascocarp: A spore producing body containing asci.
Ascogonium: The female gametangium in Ascomycotina.
Ascospores: A meiospore borne in ascus.
Ascus: A large cell in the ascocarp in which spores are developed.
Autoclave: A pressure vessel used to sterilize culture media by steam.
Basidiocarp: A spore producing body containing basidia.
Basidium (pl. basidia): The mother cell from which spores are acrogenously adjointed.
Caespitose: Aggregated in tufts.
Campanulate: Bell shaped.
Carmiplulic: The granules present in basidia having affinity to acetocarmine.
Casing: A layer of material placed on the surface of substrate to stimulate fruit body production.
Cheilocystidium (pl. cheilocystidia): A cystidium occurring at the edge of gill.
Chlamydospore: A spore with very thick membrane.
Clamps: A knot like structure.
Clavate: Club-shaped.
Compost: A mixture of substrates that undergone decomposition with the help of some microorganisms.
Context: The portion of fruit body above pores or gills.

Coralloid: The form of coral.
Crenate: Scalloped or round toothed.
Decurrent: Gills which extend downwards on the stipe.
Ecentric: Away from the centre.
Fibrilose: Covered with or composed of small fibres or hairy filaments.
Fistulose: Tubular.
Flesh: The context of the pileus of Agarics and Boletes.
Floccose: Cottony or wooly extensions from the surface.
Free (ref. to gills): Gills that do not touch the apex of stipe.
Fusiform: Spindle shaped.
Gibbous: An asymmetrical umbo or more convex on one side of the pileus than the other.
Gill: The structure resembling plates bearing hymenium of basidiomycotina.
Glabrous: Without hair.
Globose: Nearly spherical.
Granulate: Composed of oily globules.
Habitat: The place of growth.
Heterokaryotic: An individual in which genetically different nuclei are associated in the same protoplast or the same mycelium.
Heterothallic: The fungi that require the union of two compatible thalli for sexual reproduction.
Hillum: A scar showing the attachment to the sterigmata.
Hoary: Gray from fine pubescence.
Homothallic: The fungi in which sexual reproduction takes place in a single thallus that is therefore, self-compatible; non out crossing.
Hyaline: Colourless.
Hygrophanous: Water soaked in appearance.
Hymenium: A fertile layer that bears either basidia and basidiospores or asci and ascospores.
Hymenophoral trama: Trama of the lamellae or tubes.
Hypha (pl. hyphae): The unit structure of most fungi; tubular filament.
Imbricate: Overlapping one another.
Infundibulliform: Funnel shaped.
Inoculate: Transfer of an organism into a substratum.
Involute: Margin of the pileus is rolled inwards.
Irregular (ref. to trama): Pattern of hyphal arrangement not clear.
Lactiferous: Having latex.
Lamellae (gills): Plate like structures on the lower surface of pileus of mushroom bearing spores.
Lanceolate: Lance-shaped.
Liliac: Bluish.
Lumen: A central cavity bounded by a wall.
Mite: Tiny eight-legged animal with a body divided into two parts and without antennae.
Monomitic: Having one system of hyphae, the generative.
Mushroom: A fleshy, sometimes tough umbrella like basidiocarp of certain basidiomycotina and certain fleshy edible ascocarps (morel, tuber) of ascomycotina.
Mycelium: Vegetative portion of thallus of fungi composed of one or more hyphae.
Mycorrhizal: Showing a symbiotic relationship between fungus and plant.

Nematodes: Small unsegmented worms.
Non-amyloid: Hyaline or becoming yellowish with Melzer's reagent.
Obovate: Reverse of ovate.
Pallid: Somewhat pale.
Parasite: Living on another living organism and deriving food from the host.
Pasteurization: A method of selective killing of unwanted organisms by steam.
Pathogenic: Capable to cause disease.
Pliable: Plint, not firm or rigid.
Pileus: Upper portion or cap of certain types of ascocarps or basidiocarps.
Pitted: Covered with little depressions.
Pleurocystidium (pl. pleurocystidia): A cystidium occurring on the side of gill.
Pruinose: Finely powdered.
Pseudorhiza: A root like extension of the stipe, it actually acts as storage and conducting organ.
Punctuate: Having dots or spots or minute scales or granule over the surface.
Reflexed: Margin of the pileus is turned back.
Regular: Straight hyphae and not curved.
Rostrate: With a beak.
Saprophitically: Living on dead organic substance.
Sessile: Without a stipe (stalk).
Spathulate: Spatula-shaped.
Spawn: A myceliated substrate used as seed material.
Spawn run: Growing mycelia from spawn medium to substrate medium.
Species: A type of organism.
Spore: A reproductive unit (asexual).
Spore print: Mass of spores discharged from fruit body and settled on a glass plate.
Sporophore: Any structure that bears spores.
Sterigmata: A small hyphal branch or structure which supports a basidiospore.
Sterilization: The method to eliminate all organisms in a given medium or substrate by steam and pressure.
Stipe: A general term of stalk.
Stipitate: Having stipe.
Striate: Provided with fine lines.
Subculture: A culture derived from another culture.
Tomentose: Pileus or stipe covered with densly matted hairs.
Trama: The fungal tissue composing the pileus or bearing the hymenium of some basidiomycotina.
Umbo: A central elevation of the pileus.
Umbonate: Provided with a raised central region of pileus.
Ventricose: Enlarged or swollen at the middle.
Veil: Membranous structure covers a part (lamellae) of basidiocarp.
Viscid: Sticky from a tenacious coating or secretion.
Volva: A covering or a sac enclosing sporophore of agaric.
Volvate: With volva.

References

Adsule, P.G., Girija, V., Amba Dan and Tewari, R.P. (1981). A note of simple preservation of oyster mushroom (*Pleurotus sajor-caju*), *Indian J. Mushrooms*, 7(1 and 2): 2–5.

Agarwala, R.K. (1973). How to grow mushrooms? *Indian J. Mushrooms*, 17–21.

Ahlawat, O.P. and Verma, R.N. (2000–2001). Annual report 2000–01, All India Coordinated Mushroom Improvement Project, NRCM, Solan, p. 80.

Ainsworth, G.C., Sussman, A.S. and Sparrow, F.K. (Eds.) (1973). *The Fungi*, Vol. 4B, Academic Press, New York.

Alexopoulos, C.J. and Mims, C.W. (1979). *Introductory Mycology*, 3rd ed., Wiley Eastern, New Delhi, p. 632.

Alexopoulos, C.J., Mims, C.W. and Blackwell, M. (2002). *Introductory Mycology*, John Wiley & Sons, Singapore, p. 869.

Amuthan, G., Viswanathan, R., Kailappan, R. and Sreenarayanan, V.V. (1999). Studies on osmo-air drying of milky mushroom *Calocybe indica*, *Mushroom Research*, 8(2): 49–52.

Anantheswaran, S. and Sunkara, R. (1996). Use of commercial moisture absorbers to increase the shelf-life of fresh mushrooms, *Mushroom News*, 24–26.

Anderson, E.E. and Fellers, C.R. (1942). The food value of mushrooms (*Agaricus campestris)*, *Proc. Amer. Soc. Hort. Sci.*, 41: 301–303.

Arita, I. (1978). *Pholiota nameko*, pp. 475–496. In: *The Biology and Cultivation of Edible Mushrooms*, S.T. Chang and Hayes, W.A. (Eds.), Academic Press, Inc., New York, p. 819.

Arora, S., Shivhare, U.S., Ahmed, J. and Raghavan, G.S.V. (2003). Drying kinetics of *Agaricus bisporus* and *Pleurotus florida* mushrooms, *Transactions of ASAE*, 46(3): 721–724.

Arunmuganathan, T., Hemakar, A.K. and Rai, R.D. (2004). Studies on drying characteristics and effect of pretreatments on the quality of sun-dried oyster mushroom *Pleurotus florida*, *Mushroom Research*, 31(1): 35–38.

Arumuganathan, T., Rai, R.D. and Hemakar, A.K. (2003). Preparation of pickle from oyster mushroom (*Pleurotus florida*), *Processed Food Industry*, 7(3): 21–23.

Atkins, F.C. (1983). *Mushroom J.*, 125: 168–171.

Azad, K.C., Srivastava, M.P., Singh, R.C. and Sharma, P.C. (1987). Commercial preservation of mushrooms–1, A technical profile of canning and its economics, *Indian J. Mushrooms*, 12–13: 21–29.

Bahl, N. (1984). *Handbook on Mushrooms*, Oxford and IBH, New Delhi, p. 123.

Bahl, N. and Chowdhury, P.N. (1980). *Podospora faurelli*, a new competitor in the mushroom cultivation (*Volvariella volvacea*), *Curr. Sci.*, 50: 378.

Bahl, N., Ghai, S. and Prasad, D. (1981). Pest problems of button mushroom (*Agaricus bisporus*), *3rd Intl. Symp. Plant Pathol.*, New Delhi, pp. 209 (Abstr.).

Bhatt, P., Kushwaha, K.P.S. and Singh, R.P. (2007). Evaluation of different substrates and casing mixtures for production of *Calocybe Indica*, *Indian Phytopath.*, 60(1): 128–130.

Bahukhandi, D. (1990). Effect of various treatments on paddy straw on yield of some cultivated species of *Pleurotus*, *Indian Phytopath.*, 43(3): 471–472.

Bahukhandi, D. and Munjal, R.L. (1990). Studies on evolving high yielding strains of *Pleurotus sajor-caju* through hybridization, *Indian Phytopath.*, 43(1): 70–73.

Bahukhandi, D. and Sharma, R.K. (2002). Interspecific hybridization in *Pleurotus* species, *Indian Phytopath.*, 55**:** 61–66.

Bakakrishnan, B. and Nair, M.C. (1995). Production technology of oyster mushroom (*Pleurotus* sp.), In: *Advances in Horticulture,* K.L. Chadha and S.R. Sharma (Eds.), Vol. 13 (Mushroom), pp.109–119, Malhotra Publishing House, New Delhi, India.

Balazs, S. (1974). 'Stropharia growing problems in Hungary', *Rep. Veg. Crops Res. Inst.*, Keeskemet, Hungary.

Bano, Z., Nagraja, N., Vibhakar, K.V. and Kapur, O.P. (1981). Minerals and heavy metal contents in sporophores of *Pleurotus* species, *Mushroom Newsletter for Tropics*, 1(3): 6–10.

Bano, Z., Narasimhan, P. and Singh, N.S. (1975). Efficacy of some fungicides in controlling *Penicillium degitatum* Sacc., Infection during the cultivation of *Pleurotus flabellatus* (Berk. and Br.) Sacc., *Indian J. Microbiol.*, 15: 68–74.

Bano, Z. and Rajarathnam, S. (1982). *Pleurotus* mushroom as nutritious food. In: *Tropical Mushroom—Biological Nature and Cultivation Methods*, Chang, S.T. and Quimo, T.H. (Eds.). The Chinese University Press, Hong Kong, pp. 363–82.

Bano, Z., Rajarathnam, S. and Nagaraja, N. (1987). Some important studies on *Pleurotus* mushroom technology, pp. 53–64. In: *Indian Mushroom Science:* II, Proceeding of the international conference on Science and Cultivation of Edible Fungi, Kaul, T.N. and Kapur, B.M. (Eds.) Regional Research Laboratory, Council of Scientific and Industrial Research, Jammu Tawi, p. 421.

Bano, Z., Rajarathnam, S. and Shashirekha, M.N. (1992). Mushrooms—Unconventional Single Cell Protein for a Conventional Consumption, *Indian Food Packer*, 46: 20–31.

Bano, Z., Rajarathnam, S., Shashirekha, M.N. and Ghosh, P.K. (1997). *Advances in Mushroom Biology and Production*, Mushroom Society of India, NRCM, Solan, 321–337.

Bano, Z. and Srivastava, H.C. (1962). Studies in the cultivation of *Pleurotus* species on paddy straw, *Food Sci.*, **11:** 36–38.

Bano, Z., Srinivasan, K.S. and Srivastava, H.S. (1962). *Appl. Microbiology*, 11: 301–303.

Beelman, R.B. (1987). Factors influencing post harvest quality and shelf-life of fresh mushrooms, *Mushroom News.*, 35(7): 12–18.

Beelman, R.B. and Edwards, C.E. (1989). Variability in the protein content and canned product yield of four important processing strains of the cultivated mushrooms (*Agaricus bisporus*), *Mushroom News*, 37(7): 17–26.

Bels, P.J. (1976). In: *The Biology and Cultivation of Edible Mushrooms,* Chang, S.T. and Hayes, W.A. (Eds.), Academic Press, New York.

Benoit, M.A., Aprano, G.D. and Lacroix, M. (2000). Effect of gamma irradiation on phenylalanine ammonia-lyase activity, total phenolic content and respiration of mushrooms (*Agaricus bisporus*), *Journal of Agricultural and Food Chemistry*, 48(12): 6312–6316.

Besstte, A.E. (1985). Use of the mushroom tissue block rapid pitting test to detect brown blotch pathogens, *Appl. Environ. Microbiol.*, 49: 999–1000.

Beyer, D.M. (1999). Mushroom substrate preparation for white button mushroom, Vija. Wilkinson@psu.edu., pp. 1–3.

Bhandal, M.S. and Mehta, K.B. (1989). Cultivation of *Auricularia polytricha* (Mont.) Sacc. on wheat straw in India, *Mushroom Sci.*, 12: 387–393.

Bhandari, A. and Singh, R.D. (1983). Effect of some insecticides on springtail population and on the mycelial growth and sporophore yield of *Pleurotus sajor-caju* (Fr.) Singer, *Indian J. Mushrooms*, 9: 36–39.

Bhardwaj, S.C., Jandaik, C.L. and Beig, G.M. (1987). *Gliocladium virens*—a new pathogen of *Pleurotus* species, *Mush. J. Tropics*, 7: 97–100.

Bharadwaj, S.C. and Seth, P.K. (1983). Studies on green mould *Trichoderma viride* (Pers ex S.F. Gray) and its control, *Indian J. Mush.*, 9: 44–51.

Bharadwaj, S.C., Seth, P.K. and Guleria, D.S. (1989). A note on the occurrence of white plaster disease of cultivated mushroom (*Agaricus bisporous*) caused by *Scopulariopsis fimicola* a new fungus from India., *Curr. Sci.*, 58: 320–321.

Bhatt, P., Kushwaha, K.P.S. and Singh, R.P. (2007). Evaluation of different substrates and casing mixtures for production of *Calocybe indica*. *Indian Phytopath.* 60(1): 128–130.

Bisaria, R., Madan, M. and Bisaria, V.S. (1987). Mineral content of mushroom, *Pleurotus sajor-caju*, cultivated on different agrowastes. *Mushroom J. Tropics*, 7: 53–60.

Biswas, P., Sarkar, B.B., Chakraborty, D.K. and Mukherjee, N. (1983). A new report on bacterial rotting of *Pleurotus sajor-caju*, *Indian Phytopath.*, 36: 564.

Biswas, S. (1988). Physiological studies of some edible gill fungi, PhD Thesis, University of Calcutta, West Bengal, India.

Biswas, S. and Singh, N.P. (2004) Productivity of different species of oyster mushroom under Tripura condition, *Journal of Hill Research*, 17(2): 74–76.

____ (2007). Effects of climatic conditions and substrates on the production of *Pleurotus sajor-caju* in Tripura, *Environment and Ecology*, 25(1): 168–172.

____ (2008). Influence of Climatic Conditions and Cultural Practices on the Development of Fruit Bodies of *Pleurotus flabellatus* in Tripura, *Environment and Ecology*, 26(2A): 792–795.

____ (2008a). Influence of environment in cultivating edible mushroom, In: *Climate Change and Food Security*. M. Datta, N.P. Singh and D. Daschaudhuri (Eds.) pp. 267–274. New India Publishing Agency, New Delhi (India).

____ (2008b). *Mushroom Production—A Technology Bulletin for Tripura*, ICAR Research Complex for NEH Region, Tripura Centre, Lembucherra, Publication No. 39, p. 41.

Biswas, S. and Singh, N.P. (2009). Determination of seasons and substrates for paddy straw mushroom cultivation in Tripura, *Mushroom Research*, 17(2): 75–77.

____ (2009a). Evaluation of alternate substrates for milky mushroom, *J. Mycol. Pl. Pathol.*, 39(2): 355–357.

—— (2010). Agricultural residues for mushroom cultivation. In: *Sustainable Food Security*, pp. 249–260, Jain, P.K., Hansra, B.S., Chakraborty, K.S. and Kurup, J.M. (Eds), A. Mittal Publications, New Delhi, p. 401.

____ (2011). Productivity of oyster mushroom (*Pleurotus florida*) on agricultural wastes in Tripura, *Environment and Ecology*, 29(1A): 363–366.

Block, S.S., Trao, G. and Han, L. (1959). Experiments in the cultivation of *Pleurotus ostreatus*, *Mushroom Sci.*, 4: 309–325.

Bonnefons, N. De (1650). Le Jardinier Francais, Paris.

Bose, S.R. (1921). Possibilities of mushroom industry in India by cultivation, *Agric. J. India*, 16: 374–354.

Brar, D.S. and Sandhu, G.S. (1990). Damage to white button mushroom in relation to larval population of *Bradysia tritici* (Coq.), *J. Insect Sci.*, 3(2): 146–147.

Briones, G.L., Varoquaux, P., Chabroy, J., Bouquant, G., Bureau, G. and Pascat, B. (1992). Storage of common mushroom under controlled atmospheres, *International Journal of Food Science and Technology*, 27: 493–505.

Burnett, J.H. (1975). *Mycogenetics*, Wiley, New York.

Burton, K.S. (1988). Modified atmosphere packaging, *Mushroom Journal*, 183:510.

Burton, K.S. and Maher, M.J. (1991). Modified atmosphere packaging of mushrooms—review and recent developments, *Mushroom Science*, 13(2): 683–688.

Callow, E. (1831). Observations on methods now in use for artificial growth of mushrooms, with full explanation of an improved mode of culture, *Fellows*, London.

Campbell, J.D., Stothers, S., Vaisey, M. and Berck, B. (1968). Gamma radiation influence on the storage and nutritional quality of mushroom, *Journal of Food Science*, 3: 540–542.

Chadha, K.L. (1992). Mushroom Research and Development in India, *Mushroom Research*, 1: 1–12.

Chadha, K.L. and Sharma, S.R. (1995). Advances in Horticulture, *Mushroom*, Vol. 13, Malhotra Publishing House, New Delhi, p. 649.

Chakravarty, D.K. and Sarkar, B.B. (1982). Cultivation of *Pleurotus sajor-caju* on wood logs. *Indian Phytopath.* 35(2): 291–292.

Chakravarty, D.K., Sarkar, B.B., Datta, S. and Chatterjee, M.L. (1987). Bionomics and control of sciarid fly, *Lycoriella auripila* Winn. in subtropical mushroom, *Pleurotus sajor-caju* (Fr.) Sing. *Indian Mushroom Science,* 2: 146–50.

Chakravarty, D.K., Sarkar, B.B. and Kundu, B.M. (1981). Cultivation of *Calocybe indica* a tropical edible mushroom, *Curr. Sci.*, 50: 550.

Chambry (1810). Cultivation in Western Countries: Growing in Caves, by Delmas, J., 1978, pp. 251–298, In: *The Biology and Cultivation of Edible Mushrooms*, S.T. Chang and Hayes, W.A. (Eds.), Academic Press, Inc., New York, p. 819.

Chandra, A. and Purkayastha, R.P. (1977). Physiological studies on Indian mushrooms, *Trans. Br. Mycol. Soc.*, 69: 63.

Chandra, S., Singh, A.K., Bhat, M.V. and Kumar, S. (1998). Yield performance of *Pleurotus sajor-caju* on some selected substrates in North-Eastern region of India, *Mushroom Res.*, 7: 79–80.

Chandra, S., Singh, A.K. and Kumar, S. (1995). *Cultivation of oyster mushroom*, Technical Bulletin, p. 25. ICAR Research Complex for NEH Region, Umroi Road, Barapani, Meghalaya.

Chang, S.T. (1965). Cultivation of the straw mushroom in S.E. China, *World Crops*, 17: 47–49.

____ (1978). Nuclear behaviour utilizing light microscopy, pp. 35–51, In: *The Biology and Cultivation of Edible Mushrooms*, S.T. Chang and Hayes, W.A. (Eds.), Academic Press, Inc., New York, p. 819.

____ (1978a). *Volvariella volvacea*. pp. 573–603. In: *The Biology and Cultivation of Edible Mushrooms*, S.T. Chang and Hayes, W.A. (Eds.), Academic Press, Inc., New York, p. 819.

____ (1991). *Mushroom J. Tropics*, 11: 42–52.

____ (1999). World production of cultivated edible and medicinal mushrooms in 1997 with emphasis on *Lentinus edodes* (Berk.) Sing. in China, *International J. Med. Mush.*, 1: 291–300.

____ (2006). Development of culinary-medicinal mushrooms industry in China: past, present and future, *Int. J. Med. Mush.*, 8:1–17.

____ (2007). Development of the world mushroom industry and its roles in human health, pp. 1–12, In: *Mushroom Biology and Biotechnology*, R.D. Rai, S.K. Singh, M.C. Yadav and R.P. Tewari (Eds.), *Mushroom Society of India*, Solan, p. 388.

Chang, S.T. and Chu, S.S. (1969). Factors affecting spore germination of *Volvariella volvacea*, *Physiol. Plant*, 22: 734–741.

Chang, S.T. and Miles, P.G. (1987). Historical record of the early cultivation of *Lentinus* in China, *Mush. J. Tropic*, 7: 31–37.

____ (1989). *Edible Mushrooms and their Cultivation*, CRC Press, Boca Raton, pp. 345.

____ (1991). Recent trends in world production of cultivated edible mushrooms, *Mushroom* J., 504: 15–17.

____ (1997). *Mushrooms–Cultivation, Nutritional Value, Medicinal Effect and Environmental Impact*, 2nd ed., CRC Press, New York.

Chaubey, A., Dehariya, P. and Vyas, D. (2010). Yield performance of *Calocybe indica* on conventional and non-conventional substrates, *J. Mycol. Pl. Pathol*, 40(2): 176–178.

Chen, P.C. and Hou, H.H. (1978). *Tremella fuciformis*, pp. 629–643, In: *The Biology and Cultivation of Edible Mushrooms*, S.T. Chang and Hayes, W.A. (Eds.), Academic Press, Inc., New York, p. 819.

Cheng, S. and Tu, C.C. (1978). *Auricularia* sp. pp. 605–625. In: *The Biology and Cultivation of Edible Mushrooms*, S.T. Chang and Hayes, W.A. (Eds.), Academic Press, Inc., New York, p. 819.

Chi, J.H., Ha, T.M. and Kim, Y.H. (1998). Effects of packing materials for keeping freshness of *Pleurotus ostreatus* and *Flammulina velutipes*, *RDA Journal of Industrial Crop Science*, 40(2): 58–64.

Chi, J.H., Ha, T.M., Kim, Y.H. and Ju, Y.C. (1996). Effect of storage temperature and packing method for keeping freshness of fresh mushrooms, *Journal of Agricultural Science, Farm*

Management, Agricultural Engineering, Sericulture and Farm Products Utilization, 38(1): 915–921.

Chihara, G. (1992). Immunopharmacology of Lentinan, a polysaccharide isolated from *Lentinula edodes*: Its application as a host defense potentiator. *Int. J. Ori. Med.*, 17: 57–77.

Cochran, K.W. (1978). Medicinal effects, pp. 169–187, In: *The Biology and Cultivation of Edible Mushrooms*, S.T. Chang and Hayes, W.A. (Eds.), Academic Press, Inc., New York, p. 819.

Crisan, E.V. and Sands, A. (1978). Nutritional value, pp. 137–168. In: *The Biology and Cultivation of Edible Mushrooms*, S.T. Chang and Hayes, W.A. (Eds.), Academic Press, Inc., New York, p. 819.

Cui, Z.H. (2004). The production, import and export of edible mushrooms in S. Korea, The proceedings of the coordinative forum on the development of Chinese mushroom industry held in November, 2004, Zhejiang, China, 31–34.

Dar, G.M. and Seth, P.K. (1981). Studies on brown plaster mould (*Papulospora byssina* Hots.) and its control, *Indian J. Mush*, 7: 60–63.

Das, L. and Suharban, M. (1991). Fungal parasites of oyster mushroom in Kerala *Ad. Mush. Sci.*, p. 91 (Abstr.).

Dayaram (2009). Cultivation of milky white mushroom in Bihar, India, *J. Mycol. Pl. Pathol.*, 39(2): 283–285.

Dayaram, Chaurasia, S. and Patel, A. (2008). A new record of pink mould in *Pleurotus* sp. From Bihar, India, *J. Mycol. Pl. Pathol.*, 38(3): 674–675.

Declaire, J. (1978). Economics of cultivated mushrooms, pp. 727–793. In: *The Biology and Cultivation of Edible Mushrooms*, S.T. Chang and Hayes, W.A. (Eds.), Academic Press, Inc., New York, p. 819.

Delmas, J. (1978). Cultivation in western countries: Growing in caves, pp. 251–298, In: *The Biology and Cultivation of Edible Mushrooms*, S.T. Chang and Hayes, W.A. (Eds.), Academic Press, Inc., New York, p. 819.

____ (1978). The potential cultivation of various edible fungi, pp. 699–724. In: *The Biology and Cultivation of Edible Mushrooms*, S.T. Chang and Hayes, W.A. (Eds.), Academic Press, Inc., New York, p. 819.

Deshpande, A.G. and Tambane, D.V. (1982). Mushrooms—*Volvariella volvacea*. Improvement of yield, *Indian Food Packer*, 36: 56–61.

Dewangan, U.K., Thakur, M.P. and Shukla, C.S. (2007). Anti-microbial activity of culture filtrate of *Lentinula edodes* against competitor moulds, *J. Mycol. Pl. Pathol.*, 37(2): 208–209.

Dhar, B.L. (1995). Production technology of *Agaricus bitorquis*, pp. 99–107. In: *Advances in Horticulture*. K.L. Chadha and S.R. Sharma (Eds.), Malhotra Publishing House, New Delhi, India, p. 649.

____ (1997). *Advances in Mushroom Biology and Production*, R.D. Rai, B.L. Dhar and R.N. Verma (Eds.), pp. 369–378.

Dhar, B.L., Ahlawat, O.P., Dubey, J.K., Patyal, S.K. and Thakur, M. (2007). Organic farming of mushrooms—Importance, status and technology, pp. 149–165. In: *Mushroom Biology and Biotechnology*, R.D. Rai, S.K. Singh, M.C. Yadav and R.P. Tewari (Eds.), Mushroom Society of India, NRCM, Solan, p. 388.

Dieleman-van Zaayen, A. (1972). Spread, prevention and control of mushroom virus disease, *Mushroom Sci.*, 8: 131–154.

Doshi, A., Munot, J.F. and Chakravarti, B.P. (1988). Nutritional status of an edible mushroom *Calocybe indica* P and C, *Ind. J. Mycol. Pl. Pathol.*, 18: 301–302.

Doshi, A. and Sharma, S.S. (1995). Production technology of specialty mushrooms, pp. 135–154. In: *Advances in Horticulture*, Vol. 13, Mushroom, K.L. Chadha and Sharma, S.R. (Eds.), Malhotra Publishing House, New Delhi, India, p. 649.

Doshi, A., Sharma, S.S. and Trivedi, A. (1991). Problems of competitor moulds and insect—pests and their control in the beds of *Calocybe indica* P. and C, *Adv. Mush. Sci.*, pp. 57 (Abstr.).

Doshi, A. and Singh, R.D. (1985). Control of weed fungi and their effect on yield of oyster mushroom, *Pleurotus sajor-caju. Indian J. Mycol. Plant Pathol.*, 15: 269–273.

Dugger, B.M. (1901). Physiological studies with reference to the germination of certain fungus spores, *Bot. Gaz.*, 31: 38.

Eaves, C.A. (1960). A modified atmosphere system for packages of stored fruit, *J. Hort. Sci.*, 35(2): 110–117.

Edwards, R.L. (1978). Cultivation in Western Countries: Growing in house, pp. 299–336, In: *The Biology and Cultivation of Edible Mushrooms*, S.T. Chang and Hayes W.A. (Eds.), Academic Press, New York, p. 819.

Eswaran, A. and Thomas, S. (2003). Effect of various substrates and additives on sporophore yield of *Calocybe indica* and *Pleurotus* sp., *Indian J. Mushroom*, 21: 8–10.

Fermor, T.R. (1987). Bacterial diseases of edible mushrooms and their control, In: *Developments in Crop Science 10, Cultivating Edible Fungi*, Wuest, P.J., Royse, D.J. and Beelman, R.B. (Eds.), pp. 361–370, Elsevier Sci. Publi., Amsterdam.

Fletcher J.T., Hims, M.J., and Hall, R.J. (1983). The control of bubble disease and cobweb disease of mushroom with Prochloraz, *Plant Pathology*, 32: 123–131.

Fletcher, J.T., White, P.W. and Gaze, R.H. (1986). *Mushrooms: Pest and Disease Control*, Intercept Limited, Ponteland, Newcastle upon Tyne, pp. XX+ 156.

____ (1989). *Mushrooms: Pest and Disease Control*, Intercept Limited, Ponteland, Newcastle upon Tyne, p. 174.

Food and Agriculture Organization (1970). Amino Acid Content of Foods and Biological Data on Proteins, *Nutrr. Stud.*, No. 24., Food and Policy and Food Sci. Serv., Nutr. Div., Food Agric. Organ., Rome.

____ (1972). *Food Composition Table for Use in East Asia*, Food Policy and Nutr. Div., Food Agric. Organ, U.N., Rome.

Gandy, D.G., Duncan, C.W. and Edward, R.L. (1953). *Mush. Sci.*, 2: 167–174.

Gandy, D.G. and Hollings, M. (1962). Die back of mushrooms: a disease associated with a virus, *Rep. Glasshouse Crops Res. Inst.* 1961: 103–108.

Gandy, D.G. and Spencer, D.M. (1981). Fungicide evaluation for control of dry bubble caused by *Verticillium fungicola* on commercial mushroom strains. *Scientia Horticulturae*, 14: 107–115.

Ganeshan, G. (1987). A new disease of oyster mushroom by *Arthobotrys pleuroti* sp. Nov.—A first report, *Mushroom Inform.*, 4(10): 12–13.

Garasiya, N., Sharma, S.S., Gour, H.N. and Singh, R. (2007). Cultivation of jelly mushroom (*Auricularia polytricha* (Mont.) Sacc.) in Rajasthan, *India. J. Mycol. Pl. Pathol.*, 37(2): 332–335.

Garcha, H.S. (1980). *Mushroom Growing*, Punjab Agricultural University, Ludhiana.

____ (1984). Diseases of mushrooms and their control, *Indian Mushroom Science*, 1: 185–191.

Garcha, H.S., Khanna, P. and Sandhu, G.S. (1987). Status of pests in the cultivated mushroom in India. In: *Cultivating Edible Fungi*, pp. 649–665, Wuest, P.J., Royse, D.J. and Beelman, R.B. (Eds.), Elsevier, Amsterdam, p. 677.

Garcha, H.S. and Kiran, U. (1981). Studies on mushroom composts under tropical condition, *Mushroom Sci.* 11(1): 219–233.

Ghosh, R.N., Patnaik, N.C. and Tewari, I. (1967). Studies on Indian Agaricales, *Indian Phytopath*, 20: 237–242.

Gill, G.S. and Mangat, I.S. (1991). *Package of Practices for Vegetable and Fruit Crops*, Punjab Agricultural University, Ludhiana Publication, p. 100.

Gill, R.S. and Sandhu, G.S. (1994). Description and pest status of *Seira iricolor* Yosii and Ashraf (Collembia: Entomobryidae) on mushrooms in Punjab, *J. Insect Sci.* (cited by Sandhu, 1995).

Gill, W.M. (1994). Drippy Gill—An ooze disease of the cultivated mushroom *Agaricus bisporous* caused by *Pseudomonas agaraci*, Ph.D. Thesis, University of Canterburry, Christchurch, New Zealand, p. 218.

____ (1995). Bacterial diseases of *Agaricus* mushrooms, *Rep. Tottori Mycol. Inst*, 33: 34–55.

Goltapeh, E.M., Jandaik, J.N., Kapoor, J.N. and Sharma, V.P. (1989). *Cladobotryum verticillatum* – A new pathogen of Jew's ear mushroom causing cobweb disease, *Indian Phytopath*, 42: 305 (Abstr.).

Goltapeh, E.M. and Kapoor, J.N. (1990). VLP's in white button mushroom in North India, *Indian Phytopath*, 43:254 (Abstr).

Gonzalez, P. and Labarère, J. (2000). Phylogenetic relationships of *Pleurotus* species according to the sequence and secondary structure of the mitochondrial small-subunit rRNA V4, V6 and V9 domains, *Microbiology*, 146, 209–221.

Gothandapani, L., Parvathi, K. and Kennady, Z.J. (1997). Evaluation of different methods of drying on the quality of oyster mushroom (*Pleurotus* sp.), *Drying Technology*, 15(6–8): 1995–2004.

Guthrie, B.D. and Beelman, R.B. (1989). Control of bacterial deterioration in fresh washed mushrooms, *Mushroom Science*, 12(2): 689–699.

Guleria, D.S. (1976). A note on the occurrence of brown blotch of cultivated mushrooms in India, *Indian J. Mush.*, 2(1): 25.

Guleria, D.S. and Seth, P.K. (1977). Laboratory evaluation of some chemicals against *Cephalothecium* mould infecting mushroom beds, *Indian J. Mush.*, 3(1): 24–25.

Gupta, A., Raina, P.K., and Tikoo, M.L. (2004). Comparative evaluation of compost depth and density in polybag and wooden tray cultivation of white mushroom, *Mushroom Research*, 13(1): 13–16.

Gupta, Y. and Dhar, B.L. (1992). Effect of ruffling of casing layer at various intervals on flush appearance and yield of *A. bisporus*, NCMRT, Annual Report.

Gupta Y. and Dhar, B.L. (1993). Spawned casing increased mushroom yield in *Agaricus bisporus* under seasonal growing conditions in India. *Mushroom Research*, 2 : 25–28.

Gupta Y., Vijay, B. and Sohi, H.S. (1989). Spawned casing: Effect on yield of *Agaricus bisporus* (Lange) Sing., *Indian J. Mycol. Pl. Pathol.*, 19 : 225–226.

Gyurko, P. (1972). *Die Rolle der Belichlung be idem Anbau des Austernseillings* (*Pleurotus ostreatus*), *Mushroom Sci.*, 8: 461–469.

Halachmy, I. and Mannheim, C.H. (1991). Modified atmosphere packaging of fresh mushrooms, *Packaging Technology and Science*, 4: 279–286.

Han, Y.H., Veng, W.T., Chen, L.C. and Chan, S. (1981). Physiology and ecology of *Lentinus edodes* (Berk.), *Sing. Mushroom Sci.*, 11: 623–658.

Hayes, W.A. (1978). Biological Nature, pp. 191–217. In: *The Biology and Cultivation of Edible Mushrooms*, S.T. Chang and W.A. Hayes (Ed.), Academic Press, New York, p. 819.

Hayes, W.A. (1978). Nutrition, substrates, and principles of disease control, pp. 219–237, In: *The Biology and Cultivation of Edible Mushrooms*, S.T. Chang and Hayes, W.A. (Eds.), Academic Press, Inc., New York, p. 819.

Hayes, W.A. and Randle, P.E. (1969). Use of molasses as an ingredient of wheat straw mixture used for the preparation of mushroom composts. Rep. Glasshouse Crops Res. Inst. p. 142.

Hayes, W.A. and Shandilya, T.R. (1977). Casing soil and compost substrates used in artificial culture of *Agaricus bisporus*—the cultivates mushroom. *Indian J. Mycol. Pl. Pathol.*, 7: 5–10.

Hasselbach, O.E. and Musters, P. (1971). *Agaricus bitorquis* (quell.) Sacc. Een warteminned familielid van de champignon, *Champignoncultuur*, 15: 211–219.

Henze, J. (1989). Storage and transportation of *Pleurotus* mushrooms in atmospheres with high CO_2 concentrations, *Acta Horticulturae*, 258: 579–584.

Hirasawa, M.N., Shouji, T., Neta, K., Fukushima and Takada, K. (1999). Three kinds of antibacterial substances from *Lentinus edodes* (Berk.) Sing. (shiitake an edible mushroom), *Int. J. Antimicrob Agents*, 11(2): 151–157.

Ho, M.S. and Peng, J.T. (2006). Edible mushroom production in Taiwan, *Mushroom International,* (ISMS Newsletter) April issue, pp. 6–7.

Hobbs, C. (1995). *Medicinal Mushroom: An Exploration of Tradition, Healing and Culture.* Santa Cruz, C.A., Botanical Press, pp. 45–50.

Hollings, M., Stone, O.M. and Last, F.T. (1963). A virus disease of a fungus die-back of cultivated mushroom, *Endeavour*, 22: 112–117.

Hollings, M, Stone, O.M. and Atkey, P.T. (1971). Mushroom viruses, *Rep. Glassware Crops Res. Inst.*, 1967: 157–159.

Holtz, R.B. and Schisler, L.C. (1971). Lipid metabolism of *Agaricus bisporus* (Lange) Sing.: 1, Analysis of sporophore and mycelial lipids, *Lipids*, 6(3): 176–180.

Hsu, H.K. and Han, Y.H. (1981). Physiological and ecological properties and chemical control of *Mycogone perniciosa* Magn., causing wet bubble in cultivated mushroom *Agaricus bisporus*, *Mushroom Sci.*, XI: 403–425.

Huang, B.H., Yung, K.H. and Chang, S.T. (1985). The sterol composition of *Volvariella volvacea* and other edible mushrooms, *Mycologia*, 77: 959.

Huang, B.H., Yung, K.H., and Chang, S.T. (1989). Fatty acid composition of *Volvariella volvacea* and other edible mushrooms, *Mushroom Sci.*, 12(2): 533–540.

Ito, T. (1978). Cultivation of *Lentinus edodes*, pp. 461–473. In: The *Biology and Cultivation of Edible Mushrooms*, S.T. Chang and W.A. Hayes (Eds.), Academic Press, New York, p. 819.

Jana, K.K. and Purkayastha, R.P. (1982). A new record of edible *Russula* from India, *Curr. Sci.*, 51(17): 844.

Jandaik, C.L. and Kapoor, J.N. (1974). Studies on cultivation of *Pleurotus sajor-caju*. *Mush. Sci.,* 9: 667–672.

Jandaik, C.L. and Sharma, A.D. (1983). Effect of environmental parameters on the development of *Sibirinia* rot of cultivated mushrooms, *Taiwan Mushrooms*, 6(1,2): 43–47.

Jandaik, C.L., Sharma, V.P., Raina, R. (1993). Yellow blotch of *Pleurotus sajor-caju* (Fr.) Singer—a bacterial disease new to India, *Mush. Res.*, 2: 45–48.

Jarial, R.S. and Shandilya, T.R. (2004). Population of casing microflora and its relationship with production of *Agaricus bisporus*, *J. Mycol. Pl. Pathol.*, 386–390.

Jong, S.C. (1978). Conservation of the cultures, pp. 119–135. In: *The Biology and Cultivation of Edible Mushrooms*, S.T. Chang and Hayes, W.A. (Eds.), Academic Press, Inc., New York, p. 819.

Jong, S.C. and Brimingham, J.M. (1993). Medicinal and therapeutic value of the shiitake mushroom, *Ad. App. Microbiol.*, 39: 153–184.

Kachroo, J.L., Ahmad, N., Amanullah and Kaul, T.N. (1979). Experiments in the improvement of compost formulae and procedure in *Agaricus bisporus* cultivation, *The Mushroom Journal*, 82: 433–441.

Kaneda, T. and Tokuda, S. (1966). Effect of various mushroom preparations on cholesterol levels in rats, *J. Nutr.*, 90: 371–376.

Kannaiyan, S. (1974). *Edible mushrooms III*, Cultivation of *Volvariella diplasia* Far. Fac. February.

Kannaiyan, S. and Prasad, N.N. (1978). Production of paddy straw mushroom in India, *Indian Mush. Sci.*, 1: 287–293.

Kannaiyan, S. and Ramaswamy, K. (1980). *A Handbook of Edible Mushrooms,* Todays and Tomorrow's printers and publishers, New Delhi, p. 104.

Kapoor, J.N. (1989). *Mushroom Cultivation*, ICAR, New Delhi.

Kar, A. and Gupta, D.K. (2001). Osmotic dehydration characteristics of button mushrooms, *J. Food Sci. Tech.*, 38(4): 357.

Katyal, P., Khanna, P.K. and Kapoor, S. (2006). Toxicity of fungicides against *Agaricus bisporus* and its Mycopathogens. *J. Mycol. Pl. Pathol*, 36(2): 296–298.

Kaul, T.N., Kachroo, J.L. and Ahmed, N. (1978). Diseases and competitors of mushroom farms in Kashmir valley, *Indian Mush. Sci.*, 1: 193–203.

Kaur, K., Kapoor, S. and Phatela, R.P. (2004). Suitability of organic supplements in paddy straw and cotton waste compost for production of *Volvariella volvacea, Mushroom Research*, 13(1): 31–34.

Khader, V. and Pandya, B.N. (1981). Acceptability studies on weaning foods and a pickle prepared out of paddy straw mushroom (*Volvariella volvacea*) and the keeping quality of the same, *Indian Journal of Mushrooms*, 7(1 and 2): 31–36.

Kim, G.P. (1975). The toxicity of benomyl and BCM to some edible fungi and pathogenic organisms causing major disease of *Agaricus bisporus*. Research reports of the office of Rural Development, Soil Science, Fertilizer, *Plant Protection and Mycology*, Korea, 17: 137–147.

Kim, D.S. (1978). Cultivation in Asian Countries: Growing in temperate zones, pp. 345–363. In: *The Biology and Cultivation of Edible Mushrooms,* S.T. Chang and W.A. Hayes (Eds.), Academic Press, New York, p. 819.

Kindt, V. (1971). *Champignous Selbst Angebaut*, VEB Dtsch, Landwirtsch, Verlag, Berlin.

Komatsu, M. (1961). Morphological characters of the hyphae of *Lentinus edodes* (Berk) Sing., grown under the fluctuated temperatures and those during fruiting, *Rep. Tottori Mycol. Inst.*, Japan 1: 45–59.

Komatsu, M. and Goto, M. (1974). A bacterial disease of cultivated shiitake mushroom, *Lentinus edodes* (Berk.) Sing, In: *Rep. Tottori Mycol. Inst.,* Japan, 11: 69–82.

Kong, W.S. (2004). Descriptions of commercially important *Pleurotus* species, pp. 53–61, In: *Mushroom Growers Handbook 1*, Mush World.

Koorapati, A., Foley, D., Pilling, R. and Prakash, A. (2004). Electron-beam irradiation preserves the quality of white button mushroom (*Agaricus bisporus*) slices, *Journal of Food Science*, 69(1): 25–29.

Krishnamoorthy, A.S. and Muthusamy, M. (1997). Yield performance of *Calocybe indica* (P and C) on different substrates, *Mush Res*. 6(1): 29–32.

Kumar, K.R., Bano, Z., Subramanian, L. and Anandaswamy, B. (1990). Moisture sorption, storage and sensory quality studies on dehydrated mushrooms. *Indian Food Packer*, 39(5): 3–8.

Kumar, P., Pal, J. and Sharma, B.M. (2004). Cultivation of *Pleurotus sajor-caju* on different substrate combinations. *J. Mycol. Pl. Pathol*., 34(2): 322–324.

Kumari, K., Dhanda, S., Kapoor, S. and Khanna, P.K. (2008). Evaluation of *Volvariella* Species on paddy straw for yield potential and substrate degradation capacity, *J. Mycol. Pl. Pathol*, 38(2): 328–332.

Kurtzman, R.H. (1978). *Coprinus fimetarius*, pp. 393–408. In: *The Biology and Cultivation of Edible Mushrooms*, S.T. Chang and Hayes, W.A. (Eds.), Academic Press, Inc., New York, p. 819.

Lambert, E.B. (1930). The two new diseases of cultivated mushrooms. *Phytopathology*, 20: 376–384.

____ (1932). *Mushroom Growing in the United States*, U.S. Dept. Agr., Washington Circ. 251.

____ (1941). Studies on the preparation of mushroom compost. *J. Agric. Res.,* 62: 415–422.

Lee, S.C. (1596). Pen Tsao a Chinese material medica. Cited in the *Chinese Encyclopedia,* M.L. Chen (Ed.) Vol. 65, pp. 555–557, Book World Co., Taipei, 1964.

Lee, T.F. and Chang, S.T. (1975). Nutritional analysis of *Volvariella volvacea, J. Chin. Soc. Hortic. Sci*., 21: 13–20.

Leh, H.O. (1975). Bleigehalte in Pilzen, *Z. Lebensum. Unters. Forsch*., 157: 141–142.

Lelliot, R.A. and Stead, D.E. (1987). Diagnostic procedures for bacterial plant diseases. In: *Methods in Plant Pathology,* Vol. 2, Methods for the diagnosis of bacterial diseases of plants, pp. 37–131, Blackwell Scientific Publications, Oxford.

Lescane, C. (1994). Extension of mushroom (*Agaricus bisporus*) shelf life by gamma radiation, *Post Harvest Biology and Technology*, 4: 255–260.

Liang, P., Chen, K. and Lieu, H. (1987). Virus particles in sporophores of oyster mushroom, *Chinese J. Virol.*, 3: 369–375.

Liu, H.D. and Liang, P. (1986). Ultrasections of virus infected *Pleurotus ostreatus*, *Acta Microbiol. Sinica,* 26: 221–225.

Ludenberg (1754). Cultivation in Western Countries: Growing in Caves, by Delmas, J., 1978, pp. 251–298, In: *The Biology and Cultivation of Edible Mushrooms*, S.T. Chang and Hayes, W.A. (Eds.), Academic Press, New York, p. 819.

Mallesha, B.C. and Shetty, K.S. (1988). A new brown spot disease of oyster mushroom caused by *Pseudomonas stutzeri*, *Curr. Sci.*, 57: 1190–1192.

Mamatha (1998). In: *Economics of Oyster Mushroom Cultivation* (Upadhyay, 2010). Training manual of training programme on mushroom cultivation technology. Organized by Directorate of Mushroom Research, Solan, Himachal Pradesh in collaboration with Directorate of Horticulture and Soil Conservation, Government of Tripura at Horticulture Research Complex, Nagicherra, Dated 5th October, 2010 to 7th October 2010.

Mantel, E.F.K., Agarwala, R.K. and Seth, P.K. (1972). *A Guide to Mushroom Cultivation*, Farm Bull. 2. Farm Information Unit, Directorate of Extension, Ministry of Agriculture, New Delhi.

Markhija, A.N. and Bailley, J.M. (1981). Identification of the antiplatelet substance in Chinese black tree fungus (letter), *N. Engl. J. Med.*, 304: 175.

Mc Connel, J.E. and Esselen, W.B. (1947). Carbohydrates in cultivated mushrooms, *Food Res.*, 12: 118–121.

Meena, R.D. and Sharma, S.S. (2005). Evaluation of substrates for cultivation of *Lentinula edodes*, *J. Mycol. Pl. Pathol.*, 35(2): 219–221.

Minamide, T., Habu, T. and Ogata, K. (1980). Effect of storage temperature on keeping freshness of mushrooms after harvest, *J. Jpn. Soc. Fd. Sci.Tech.*, 27(6): 281–287.

Mizuno, T. (1999). The extraction and development of antitumor active polysaccharides from medicinal mushroom in Japan, *Int. J. Med. Mush.*, 1: 9–29.

Morris, E., Doyle, O. and Clancy, K.J. (1995). A profile of *Trichoderma* species II. Mushroom growing units. *Mush. Sci.,* 14(2): 619–625.

Mudahar, G.S. and Brains, G.S. (1982). Pre-treatment effect on quality of dehydrated *Agaricus bisporus* mushrooms, *Indian Food Packer*, 28(5): 19–27.

Munjal, R.L. (1975). Cultivation of paddy straw mushroom, *Indian J. Mush.*, 1: 32–38.

Munjal, R.L. and Anand, J.C. (1979). *Mushroom Cultivation*, Directorate of Extension, Ministry of Agriculture and Irrigation, New Delhi.

Munjal, R.L. and Seth, P.K. (1974). Brown plaster mould—A new disease in white button mushroom, *Indian Hort.*, 18(4): 13.

Nakai, Y. and Ushiyama, R. (1974). Fine structure of shiitake mushroom, *Lentinus edodes* (Berk) Sing. I. Scanning electron microscopy on basidia and basidiospores, *Rep. Tottori Mycol. Inst.*, Japan, 11: 1–6.

Nakamura, K. (1981). *Mushroom Cultivation in Japan*, Asaki Publication House, Japan.

Nehru, C., Kumar, V.J.F., Maheshwari, C.U. and Gothandapani, L. (1995). Solar drying characteristics of oyster mushroom, *Mushroom Research*, 4(1): 27–30, 1995.

Nott, H.M. (1989). A study of ginger blotch disease of mushrooms, M.Sc. Thesis, University of Canterbury, Christchurch, New Zealand, p. 82.

Pilat, A. (1954). *Mushrooms*, H.W. Bijl, Amsterdam (cited in Crisan and Sands, 1978).

Pizer, N.H. (1937). Investigations into the environment and nutrition of the cultivated mushroom *Psalliota campestris*: 1. Some properties of composts in relation to the growth of the mycelium, *J. Agric. Sci.*, 27: 349.

Pizer, N.H. and Thompson, A.J. (1938). Investigations into the environment and nutrition of the cultivated mushroom (*Psalliota campestris*): II. The effect of calcium and phosphate on growth and productivity. *J. Agric. Sci.*, 28: 604–617.

Prakash, T.N. and Tejeswini, 1991. Prospects of mushroom cultivation in India. *Indian Mushroom Proc*, pp. 259–263, National Symposium on Mushrooms (1991), Thiruvananthapuram.

Promod, R., Bal Krishnan, B. and Das, L. (2004). Evaluation of different substrates for cultivation of paddy straw mushroom *Volvariella volvacea. Mushroom Research*, 13(1): 29–30.

Pruthi, J.S., Manan, J.K., Raina, B.L. and Teotia, M.S. (1984). Improvement in whiteness and extension of shelf life of fresh and processed mushrooms (*Agaricus bisporus* and *Volvariella volvacea*), *Indian Food Packer*, 38(2): 55–63.

Purkayastha, R.P., Biswas, S. and Das, A.K. (1980). Some experimental observations on the cultivation of *Volvariella volvacea* in the plains of West Bengal, *Indian Journal of Mushrooms*, 6: 1–9.

____ (1981a). Factors affecting production of paddy straw mushroom *(Volvariella volvacea*), *Indian Journal of Mushrooms*, 7: 26–30.

Purkayastha, R.P. and Chandra, A. (1976). A new technique for *in vitro* production of *Calocybe indica*—an edible mushroom of India, *The Mushroom Journal*, 40: 1–2.

____ (1985). *Manual of Indian Edible Mushrooms*, Today and Tomorrow Printers and Publications, New Delhi, p. 267.

Purkayastha, R.P. and Das, A.K. (1991). Regulation of sporophore production of paddy straw mushroom and the competitor *Coprinus* by cultural practices, pp. 153–159. In: *Indian Mushrooms*, Nair, M.C. (Ed.), Kerala Agricultural University, Vellanikkara.

Purkayastha, R.P., Das, A.K. and Biswas, S. (1981b). *Cultural Practices and Seasonal Conditions Affecting Production of Paddy Straw Mushroom,* The Taiwan Mushrooms, 5: 10–16.

Purkayastha, R.P., Mandal, T. and Jana, K.K. (1981). An improved method of cultivation of *Calocybe indica* an edible white mushroom, *Ind. J. Mushrooms*, 7: 3–10.

Purkayastha, R.P. and Nayak, D. (1977). Effect of substrates and spawn density on the production and protein content of *Calocybe indica*, *The Mushroom Journal*, 52: 132–138.

____ (1979). *A new method of cultivation of Calocybe indica: An edible summer mushroom*, The Taiwan Mushrooms, 3(1): 14–18.

____ (1981). Development of cultivation method and analyses of protein of a promising edible mushroom *Calocybe indica* P and C, *Mushroom Science*, 9: 697–713.

Puschel, J. (1969). Der Riesentrushling, ein neur Kulturpilz. Dtsch. Gartner Post No. Champignonbeilage No. 19.

Quimeo, T.H., Chang, S.T. and Royse, D.J. (1990). *Technical Guidelines for Mushroom Growing in the Tropics*, FAO, Rome.

Rai, R.D. (1995). Nutritional and medicinal values of mushrooms. pp. 537–551. In: *Advances in Horticulture*, Vol. 3, *Mushroom*, Chadha, K.L. and Sharma, S.R. (Eds.) Malhotra Publishing House, New Delhi, p. 649.

Rai, R.D. and Arumuganathan, T. (2008). *Post Harvest Technology of Mushrooms*, Technical Bulletin, National Research Centre for Mushroom, Chambaghat, Solan, p. 84.

Rai, R.D., Saxena, S., Upadhyay, R.C. and Sohi, H.S. (1988). Comparative nutritional value of various *Pleurotus* sp. grown under identical conditions, *Mushroom J. Tropics*, 8: 93–98.

Rai, R.D. and Sohi, H.S. (1988). How protein rich are mushrooms, *Indian Hort.*, 33(2): 2–3.

Rainey, P.B. and Cole, A.L.J. (1988). A new bacterial disease of the cultivated mushroom *Agaricus bisporus*, *Trans. Brit. Mycol. Soc.*, 90: 122–125.

Rama, V. and John, P.J. (2000). Effects of methods of drying and pretreatments on quality of dehydrated mushroom, *Indian Food Packer*, 54(5): 59–64.

Ramanathan, M., Sreenarayanan, V.V. and Swaminathan, K.R. (1992). Controlled atmosphere storage of mushroom, *Traction*, 2(2): 63–67.

Ramsbottom, J. (1953). Mushrooms and toadstools, *Proc. Ntr. Soc.*, 12: 39–44.

Rangad, C.O. and Jandaik, C.L. (1980). Cultivation of *Flammulina velutipes* and *Agrocybe aegerita* in India, *Indian J. Mushroom*, 6: 48–52.

Rangaswami, G. (1978). *Volvariella*—The edible paddy straw Mushroom, *Indian Mush. Sci.*, 1: 267–276.

Rath, G.C. (1962). The genus *Volvariella* in Lucknow, *J. Indian Bot. Soc.*, 41: 524–530.

Rawal, P. and Sharma, S.S. (2010). Evaluation of substrate supplements for production of black ear mushroom Auricularia polytricha, *J. Mycol. Pl. Pathol.*, 40(3): 405–407.

Raychaudhury, S.P. (1978). Virus diseases of mushrooms, *Indian Mush Sci.*, 1: 205–214.

Richardson, P.N. (1993). Stipe necrosis of cultivated mushroom *Agaricus bisporus*. *Trans. Brit. Mycol. Soc.*, 90: 122–125.

Romaine, C.P. and Schlagnhaufer, B. (1989). Prevalence of double stranded RNAs in healthy and La France disease affected basidiocarps of Agaricus bisporus, *Mycologia*, 81: 822–825.

Roy, A., Sur, C.K. and Samajpati, N. (1978). Edible mushrooms of West Bengal, III, *Volvariella, Indian Agric.*, (22), 61–62.

Roy, M.K., Chatterjee, S.R., Bahukhandi, D., Sharma, R. and Philips, A.S. (2000). Gamma radiation in increasing productivity of *Agaricus bisporus* and *Pleurotus sajor-caju* and enhancing storage life of *P. sajor-caju*, *Journal of Food Science and Technology*, 37(1): 83–86.

Roy, M.N. and Bahl, N. (1984). Gamma radiation for preservation of *Agaricus bisporus*. *Mushroom Journal*, 136: 124–125.

Roy, S., Anantheswaran, R.C. and Beelman, R.B. (1995). Fresh mushroom quality as affected by modified atmosphere packaging, *Journal of Food Science*, 60(2): 334–340.

Roy, S., Anantheswaran, R.C. and Beelman, R.B. (1996). Modified atmosphere and modified humidity packaging of fresh mushrooms, *Journal of Food Science*, 61(2): 391–397.

Royse, D.J. (2003). *Cultivation of Oyster Mushroom,* Information and Communication Technologies, College of Agricultural Sciences, Pennsylvania State University.

Royse, D.J., Schisler, L.C. and Diehle, D.A. (1985). Consumption, production, cultivation of the shiitake mushroom, *Interdisciplinary Sci. Rev.*, 10: 329–338.

Samota, R.K and Sharma, S.S. (2006). Nutritional and physical parameters required for improved strains of Shiitake (*Lentinula edodes*), *J. Mycol. Pl. Pathol.*, 36(2): 299–301.

Samota, R.K. and Sharma, S.S. (2008). Development of new isolates of shiitake mushroom through hyphal anastomosis of single spore cultures, *Indian Phytopath*, 61(2): 232–237.

Samp, R.J. and Phelps, J.K. (1986). Observations of the effects of growing temperature on first break mushrooms, pp. 265–270. In: *Cultivating Edible Fungi*, P.J. West, D.J. Royse and R.B. Beelman (Eds.) Elsevier, p. 677.

San Antonio, J.P. and Fordyce, C., Jr. (1972). Cultivation of paddy straw mushroom, *Volvariella volvacea* (Bull ex Fr.), *Sing. Hort Science*, 7: 461–464.

San Antonio, J.P. and Thomas, R.L. (1972). Carbon dioxide stimulation of hyphal growth of the cultivated mushroom *Agaricus bisporus* (Lange), *Sing. Mush. Sci.*, 623–629.

Sandhu, G.S. (1995). Management of mushroom insect pests, pp. 239–260. In: *Advances in Horticulture*. Vol. 13. *Mushroom*, K.L. Chadha and S.R. Sharma (Eds.), Malhotra Publishing House, New Delhi. p. 649.

Sandhu, G.S. and Arora, P.K. (1990). Studies on mechanical and chemical control of mushroom flies—important pests of white button mushroom in the Punjab. *J. Insect Sci.*, 3: 92–96.

Sandhu, G.S. and Bhattal, D.S. (1986). Recognition of adult and immature stages of phorid fly, *Megaselia sandhui* Disney (Diptera: Phoridae), *Mushroom News l. Trop.*, 6(4): 16–19.

Sandhu, G.S. and Brar, D.S. (1980). Biology of sciarid fly, *Bradysia tritici* (Coq.) (Diptera: Sciaridae) infesting mushroom, *Indian J. Mushrooms*, 6: 53–63.

Sangeetha, G., Prakasam, V. and Krishnamoorthy, A.S. (2008). Jaggery enhances yield of paddy straw mushroom, *Volvariella volvacea*, *J. Mycol. Pl. Pathol*, 38(2): 295–297.

Saray, T., Balla, C., Horti, K. and Sass, P. (1994). The importance of packaging and modified atmosphere in maintaining the quality of cultivated mushrooms (Agaricus bisporus L.) stored in chill chain, *Acta Horticulturae*, 368: 322–326.

Sawada (1965). *Todai Enshurin Hokoku,* 59: 33 (cited in Crisan and Sands, 1978).

Saxena, S. and Rai, R.D. (1988). Storage of button mushrooms (*Agaricus bisporus*), The effect of temperature, perforation of packs and pretreatment with potassium metabisulphite, *Mushroom J. Tropics*, 8: 15–22.

___ (1990). Post Harvest Technology of Mushrooms, *Technical Bulletin*, No. 2, NRCM, Solan.

Schanel, L. (1970). Effect of carbon dioxide on the growth of wood-decaying fungi, *Folia Biol.* (Prague), 11: 67–72.

Schisler, L.C., Sinden, J.W. and Singel, E.M. (1963). Transmission of a virus disease of mushrooms by infected spores, *Phytopathology*, 53: 888.

Seth, P.K. (1977). Pathogens and competitor of *Agaricus bisporus* and their control, *Indian J. Mush.*, 3: 31–40.

Seth, P.K. (1978). Contributions of mushroom research at Solan Centre, *Indian Mushroom Science*, 1: 67.

Seth, P.K. and Bhardwaj, S.C. (1989). Studies on Vert-de gris caused by *Myceliophthora lutea* Cost. *on Agaricus bisporu*s and its control, *Mush. Sci.*, 12(2): 725–733.

Seth, P.K. and Dar, G.M. (1989). Studies on *Cladobotryum dendroides* (Bull. Merat) W. Gams et Hozzem, causing cobweb disease of *Agaricus bisporus* (Lange) Singer and its control, *Mush. Sci.*, 12(2): 711–723.

Seth, P.K., Kumar, S. and Sandilya, T.R. (1973). Combating dry bubble of mushrooms, *Indian Hortic.*, 18: 17–18.

Seth, P.K. and Munjal, R.L. (1981). Studies on *Lilliputia rufula* (Berk and Br.) Huges and its control, *Mush. Sci.*, 11(2): 427–441.

Seth, P.K. and Shandilya, T.R. (1975). Effect on different quantities of super phosphate on the productivity of *Agaricus bisporus*, *Indian J. Mushrooms*, 1(2): 10–12.

Shandilya, T.R. (1976). *Indian J. Mushrooms*, 2: 43–46.

Shandilya, T.R. and Hayes, W.A. (1978). Use of different substrates in mushroom compost and their effect on productivity of *Agaricus bisporus*, *Indian Mushroom Sci.*, 1: 100–129.

Shandilya, T.R., Seth, P.K. and Munjal, R.L. (1975). Combating insect pests of mushroom *Agaricus bisporus* (Lange) Sing, *Indian J. Mushrooms*, 1(2): 13–15.

Sharma, A.D. and Jandaik, C.L. (1980). Studies on some competitors of two edible fungi, *Indian J. Mush.*, 6: 30–35.

____ (1981). Some weed fungi on *Pleurotus* beds, *Indian J. Mush.*, 7: 54–55.

____ (1983). Some preliminary observation on the occurrence of sibirinia rot of cultivated mushrooms in India and its control, *Taiwan Mushrooms*, 6(1, 2): 38–42.

____ (1987). Occurrence of brow rot—A new disease of *Pleurotus* beds, *Indian Mush. Sci.*, 2: 123–125.

Sharma, P.D. (2004). *The Fungi*, Rastogi Publications, India, p. 540.

Sharma, S.R. (1995). Management of mushroom diseases, pp. 195–238. In: *Advances in Horticulture*. Vol. 13, *Mushroom,* K.L. Chadha and S.R. Sharma (Eds.), Malhotra Publishing House, New Delhi. p. 649.

Sharma, S.R. and Kumar, S. (2000). Studies on wet bubble disease of white button mushroom (*Agaricus bisporus*) caused by *Mycogone perniciosa*, *Mush. Sci.* XV: 569–575.

Sharma, S.R., Kumar, S. and Sharma, V.P. (2006). Physiological requirement for cultivation of Malaysian strain of shiitake, *Lentinula edodes*, *J. Mycol. Pl. Pathol*, 36: 149–152.

Sharma, V.P., Sharma, S.R. and Kumar, S. (2006a). Comparative studies on substrate treatment methods for cultivation of *Calocybe indica*, *J. Mycol. Pl. Pathol.*, 36: 145–148.

____ (2006b). Effect of supplementation and cultivation containers on the productivity of Flammulina velutipes, *Mushroom Research*, 15(2): 129–134.

Sharma, V.P., Kumar, S. and Tewari, R.P. (2009). Flammulina velutipes: The Culinary Medicinal Winter Mushroom, *Tech. bull.*, Directorate of Mushroom Research, Chambaghat, Solan, p. 53.

Shymanski, J. (1991). *Diseases of Oyster Mushrooms*. Mikolgia I Fitolpathologia, 25: 518–523.

Siddique, B., Gogoi, R. and Puzari, K.C. (2004). Contaminant microflora of oyster mushroom beds and their effect on yield under humid climate, *J. Mycol. Pl. Pathol.*, 34(2): 329–332.

Sinden, J.W. and Hauser, E. (1950). The short method of composting, *Mushroom Sci.* 1: 52–99.

____ (1953). The nature of the composting process and its relation to short composting, *Mushroom Sci.*, 2: 123–150.

Singer, R. (1975). The Agaricales in Modern Taxonomy, 3rd edition, Cramer.

Singh, H. (1983). *Mushroom Growing in India*, Sterling Publishers, New Delhi, India

Singh, C. and Sharma, V.P. (2001). Control of wet bubble disease, *Plant Dis. Res.*, 16: 225–229.

Singh, M., Kamal, S. and Wakchaure, G.C. (2011). Earning more through exporting mushrooms. *Indian Hort.*, 56(3): 41–43.

Singh, M., Singh, A.K. and Gautam, R.K. (2007). Effect of casing and supplementation on yield of milky mushroom (*Calocybe indica*), *Indian Phytopath.*, 60(2): 191–193.

Singh, R.P. and Saxena, H.K. (1987). Studies on undesirable fungi coming up during cultivation of *Pleurotus sajor-caju* (Fr.) Sing. mushroom, *Indian Mush. Sci.*, 2: 126–128.

Singh, S., Kumar, C.G. and Singh, S. (1995). Production, processing and consumption patterns of mushrooms, *Indian Food Industry*, 14(6): 38–46.

Smith, A.H. (1978). Morphology and classification, pp. 3–34. In: *The Biology and Cultivation of Edible Mushrooms*, S.T. Chang and Hayes, W.A. (Eds.), Academic Press, New York, p. 819.

Sohi, H.S. (1986). Diseases and competitor moulds associated with mushroom culture and their control, N.C.M.R.T., Chambaghat, Solan, *Ext. Bull.*, No. 2, p. 12.

____ (1988). Mushroom culture in India—Recent research findings, *Indian Phytopath.*, 41(3): 313–326.

____ (1988). Diseases of white button mushroom (*Agaricus bisporus*) in India and their control, *Indian J. Mocol. Pl. Pathol.*, 18: 10–18.

Sohi, H.S., Seth, P.K. and Kumar, S. (1965). Occurrence of truffle disease (*Pseudibalsamia microspora* Diel and Lambert) in mushroom beds, *Curr. Sci.*, 31: 488–489.

Sohi, H.S. and Upadhyay, R.C. (1986). Occurrence of *Cladobotryum variospermum* (Link) Hughes on Polyporus fungi under natural conditions, *Curr. Sci.*, 55: 1037–1038.

____ (1989). New and Noteworthy disease problems of edible mushrooms in India, *Mush. Sci.*, 12: 611–614.

Sohliya, D.S.R., Sarmah, D.K., Gogoi, R., Puzari, K.C. and Chelleng, A. (2010). Preservation and packaging of milky mushroom, *J. Mycol. Pl. Pathol.*, 40(4): 575–580.

Staden, O.L. (1967). Radiation preservation of fresh mushroom. *Mushroom Science*, 6: 457–461.

Stamets, P. (1999). Mycomedicinals: An informational booklet on medicinal mushrooms *Mycomedia*, Olympia, W.A.: pp. 232–235.

Stanek, M. (1974). State of production and research of edible fungi, *Int. Symp. Czech. Mycol. Soc.*, Prague.

Stanek, M. and Vojtechovska, V. (1972). Test on the use of benomyl in mushroom cultivation. *Mykologicky Sbornik*, 9: 17–23.

Sugimori, T., Oyama, Y. and Omichi, T. (1971). Studies on basidiomycetes. 1. Productions of mycelium and fruiting body from non-carbohydrate organic substances, *J. Ferment. Technol.*, 49: 435–446.

Sukla, P.K. (2007). Effect of casing of milky mushroom (*Calocybe indica* P and C). *Indian Phytopath.*, 60(4): 537–539.

____ (2008). Effect of soil manure ratio of casing soil on crop duration and yield of milky mushroom (*Calocybe indica*), *J. Mycol. Pl. Pathol.*, 38(1): 47–50.

Suman, B.C. (2005). Evaluation of single spore isolates of *Agaricus bisporus* (Lange.) Sing., *J. Mycol. Pl. Pathol.*, 35(1): 103–105.

Suman, Y.S. and Sharma, B.K. (1999). Cultivation of *Volvariella volvacea* and *Pleurotus flabellatus* in submontane and low hills of Himachal Pradesh, *J. Mycol. Pl. Pathol.*, 29: 250–251

Suman, B.C. and Sharma, V.P. (2005). *Mushroom Cultivation and Uses*, Agrobios, India, p. 349.

Sumi, M. (1933). On the ergosterin content of various edible mushrooms in Japan, *Sci. Pap. Inst. Phys. Chem. Res.* Japan, 20: 254–258.

Suyama, K. and Fujii, H. (1993). Bacterial disease occurred on cultivated mushroom in Japan, *J. Agric. Sci.*, Tokyo Nogyo Daigaku, 38: 35–50 (Japanese).

Szudyga, K. (1973). *Pierscieniak*, PWRiL, Warszawa.

____ (1978). *Stropharia rugoso-annulata*, pp. 559–571. In: *The Biology and Cultivation of Edible Mushrooms*, S.T. Chang and Hayes, W.A. (Eds.), Academic Press, New York, p. 819.

Suzuki, S. and Oshima, S. (1976). Influence of shiitake (*Lentinus edodes*) on human serum cholesterol, *Mushroom Sci.* 9(1): 463–469.

Tandon, G., Sharma, V.P. and Jandaik, C.L. (2006). Evaluation of different casing materials for *Calocybe indica* cultivation, *Mush Res.*, 15(1): 35–37.

Tanga, A.Q. (1974). Weight loss in mushroom during canning, *Indian Mushroom Science*, 1: 225–232.

Tano, K., Arul, J., Doyon, G. and Castaigne, F. (1999). Atmospheric composition and quality of fresh mushrooms in modified atmosphere packages as affected by storage temperature abuse, *Journal of Food Science*, 64(6): 1073–1077.

Tewari, R.P. (1987). False truffle *Diehliomyces microsporus* as competitor in mushroom beds-a review, *Indian Mushroom Science* (CSIR, New Delhi), 2: 133–135.

Tewari, R.P. (2010). Economics of milky mushroom cultivation, In: *Training Manual of Training Programme on Mushroom Cultivation Technology*, Organized by Directorate of Mushroom Research, Solan, Himachal Pradesh in collaboration with Directorate of Horticulture and Soil Conservation, Government of Tripura, at Horticulture Research Complex, Nagicherra, Dated 5th October 2010 to 7th October 2010.

Tewari, R.P. and Singh, S.J. (1984). Mushroom virus disease in India, *Mushroom J.*, 142: 354–355.

Thakur, K. and Sharma, S.R. (1992). Substrate and supplementations for the cultivation of shiitake, *Lentinus edodes* (Berk.) Sing, *Mush. Inform.* 9(7/8): 7–10.

Thapa, C.D. and Seth, P.K. (1982). Mushroom mites and their control, *Indian J. Mushrooms*, 8: 45–52.

____ (1983). The tiny pests of mushrooms and their control, *Indian J. Mushroom*, 9: 40–43.

____ (1987). Occurrence of bacterial blotch of mushrooms (*Agaricus* sp.) caused by *Pseudomonas tolaasii* Paine in India and its control, *Indian Mushroom Sci.*, 2: 129–131.

Thapa, C.D., Seth, P.K. and Pal, J. (1979). Occurrence of olive green mould (*Chaetomium globosum* Kunze ex. Steudel) in mushroom beds and its control, *Indian J. Mush.*, 5: 9–13.

Thayumanavan, B. and Manickam, A. (1980). Protein quality of the sporophore of the fungus *Pleurotus sajor-caju* (Fr.) Singer, *Indian J. Nutr. Dietet.*, 17: 140–142.

Thomas, K.M., Ramakrishnan, T.S. and Narsimhan, I.L. (1943). Paddy Straw Mushroom, *Madras Agric. J.*, 31: 57–59.

Tirkey, B.K., Kumar, M. and Dayaram (2009). Effect of different casing on yield of *Agaricus bisporus* and *A. bitorquis* in Bihar, *Indian. J. Mycol. Pl. Pathol.*, 39(3): 471–474.

Tokimoto, K. and Komatsu, M. (1978). Biological nature of *Lentinus edodes*, pp. 445–459. *The Biology and Cultivation of Edible Mushrooms*, In: S.T. Chang and W.A. Hayes (Eds.), Academic Press, New York, p. 819.

Tominaga, Y. (1978). *Tricholoma matsuke*, pp. 683–697. In: *The Biology and Cultivation of Edible Mushrooms*, S.T. Chang and Hayes, W.A. (Eds.), Academic Press, New York, p. 819.

Tonomura, H. (1978). *Flammulina velutipes*, pp. 409–421, In: *The Biology and Cultivation of Edible Mushrooms*, S.T. Chang and Hayes, W.A. (Eds.), Academic Press, New York, p. 819.

Tourneforte, J. de. (1707) Observations sur la naissance et sur la culture deschampignons. *Memories de l'Academie Royale des Science*, 58–66.

Treschow, C. (1944). Nutrition of the cultivated mushroom, *Dansk. Bot. Ark.*, 11: 1.

Trivedi, A., Sharma, S.S. and Doshi, A. (1991). Cultivation of *Calocybe indica* under semi-arid conditions. *Indian Mushrooms*, 166–169.

Trivedi, A., Doshi, A. and Sharma, S.S. (1994). Uptake and translocation of chemicals in fruit bodies of *Calocybe indica*, *Mush. Res.*, 3: 95.

Tsuneda, A., Suyama, K., Murakami, S. and Ohira, I. (1995). Occurrence of *Pseudomonas tolaasii* on fruiting bodies of *Lentinula edodes* formed on *Quercus* logs, *Mycoscience*, 36: 283–288.

Tu, C.C. (1981). Cultivation of *Tremella fuciformis* Berk, In: *Taiwan with Special References to Cooking Techniques*, *Taiwan Mushrooms*, 5(1): 17–27.

Tuley, L. (1996). Swell time for dehydrated vegetables, *Intern. Food Ingredients*, 4(1): 23–27.

Tunney, J. (1980). Guidelines and procedure for making bulk pasteurized compost. Report submitted by Mr. Tunney to compost mother unit HDMC. Chambaghat 173213, Solan (H.P.), India.

Upadhyay, R.C. (1990). *Cultivation of Oyster Mushroom. Tech. Bull.* 1, NCMRT, Solan.

____ (2010). Economics of oyster mushroom cultivation, In: *Training Manual of Training Programme on Mushroom Cultivation Technology*. Organized by Directorate of Mushroom Research, Solan, Himachal Pradesh in collaboration with Directorate of Horticulture and Soil Conservation, Government of Tripura, at Horticulture Research Complex, Nagicherra, Dated 5th October 2010 to 7th October 2010.

Upadhyay, R.C. and Sohi, H.S. (1989). Natural occurrence of *Stropharia rugoso-annulata* Farlow apud Murill in Himachal Pradesh (India) and its artificial cultivation, *Mush. Sci.*, 12: 509–516.

Upadhyay, R.C., Sohi, H.S. and Vijay, B. (1987). *Cladobotryum apiculatum*, a new mycoparasite of *Pleurotus* beds, *Indian Phytopath.*, 40: 294 (Abstr.)

Van de Geijin, J. (1977). The control of bubble (*Verticillium fungicola* and *Mycogone perniciosa*), *Champignoncultuur*, 21: 197, 199, 201.

Van lier, J.J.C., Van Ginkel, J.T., Straatsma, G., Gerrits, J.P.G. and Van Griensven, L.J.L.D. (1994). Composting of mushroom substrate in a fermentation tunnel: compost parameters and a mathematical model, *Netherlands Journal of Agricultural Sciences*, 42(4): 271–292.

Van Zaayen, A. (1976). Immunity of strains of *Agaricus bitorquis* to mushroom virus disease, *Neth. J. Plant Path.,* 82: 121–131.

Van Zaayen, A. and Van Andrichem, J.C.J. (1982). Prochloraz for control of fungal pathogens of cultivated mushrooms. *Neth. J. Plants Pathol*., 88: 203–213.

Van Zaayen, A. and Van der Pol-Luiten, B. (1979). Heat resistance, some biological aspects and prevention of false truffles (*Diehliomyces microsporus*). *Mush. Sci*., 10(2): 319–336.

Van Zaayen, A. and Rutigens, A.J. (1981). Thermal death points for two *Agaricus* species and for the spores of some major pathogens, *Mushroom Science* XI: 393–402.

Van Zaayen, A. and Temmink, J.H.M. (1968). A virus disease of cultivated mushrooms in Netherlands, *Neth. J. Plant Pathol*., 74: 48–51.

Vedder, P.J.C. (1978). Cultivation, pp. 377–392, In: *The Biology and Cultivation of Edible Mushrooms*, S.T. Chang. and Hayes W.A. (Eds.), Academic Press, New York, p. 819.

Verma, R.N. (1980). *Cultivation of Oyster Mushroom in the North Eastern Hill States*, Extension Bulletin, ICAR Research Complex for NEH Region, Shillong.

____ (1985). *A New Technology of Pleurotus Cultivation*, ICAR Research Complex for NEH Region, Newsletter, 8(2): 9–11.

____ (2002a). *Recent Advances in the Cultivation Technology of Edible Mushrooms,* R.N. Verma and B. Vijay (Eds.), NRCM Solan, pp. 1–10.

____ (2002b). Cultivation of paddy straw mushroom (*Volvariella* sp.). In: *Recent Advances in the Cultivation Technology of Edible Mushrooms*. R.N. Verma and B. Vijay (Eds.), NRCM, Solan, pp. 221–228.

Verma, S. and Rai, R.D. (2005). *Mushroom Recipes*, NRCM, Solan, p. 73.

Vijay, B. (2006). Indoor composting for button mushroom cultivation, *Mushroom Research*, 15(1): 23–27.

____ (2010). Economics of button mushroom cultivation in controlled and seasonal growing, In: *Training Manual of Training Programme on Mushroom Cultivation Technology.* Organized by Directorate of Mushroom Research, Solan, Himachal Pradesh in collaboration with Directorate of Horticulture and Soil Conservation, Government of Tripura, at Horticulture Research Complex, Nagicherra, Dated 5th October 2010 to 7th October 2010.

Vijay, B. and Gupta, Y. (1995) Production technology of *Agaricus bisporus*, pp. 63–98. In: *Advances in Horticulture,* Vol. 13, K.L. Chadha and S.R. Sharma (Eds.), Malhotra Publishing House, New Delhi, p. 649.

Vijay, B., Gupta, Y. and Sharma, S.R. (1993). *Sepedonium maheshwarianum* – A new competitor of *Agaricus bisporus, J. Mycol. Pl. Pathol*., 23:121.

Vijay B. and Sohi, H.S. (1987). Cultivation of oyster mushroom *Pleurotus sajor-caju* (Fr.) Singer on chemically sterilized wheat straw, *Mush J. Tropics*., 7: 67–75.

____ (1989). Fungal competitors of *Pleurotus sajor-caju* (Fr.) Sing, *Mush. J. Tropics.*

Wahid, M. and Kovacs, E. (1980). Shelf-life extension of mushrooms (*Agaricus bisporus*) by gamma radiation, *Acta Alimentaria*, 9: 357–359.

Wang, H.H. and Cheng, T.F. (1978). Acid preservation of mushroom, *Mushroom Science*, 10(2): 767–775.

Wasser, S.P. and Weis, A.L. (1999). Medicinal properties of substances occurring in higher basidiomycetes mushrooms, Current perspectives (Review), *Int. J. Med. Mushr.*, 1: 31–62.

Watling, R. and Gregory, N.M. (1980). *Larger Fungi from Kashmir*, Nova Hedwigia, 32: 493–564.

Wong, W.C., Fletcher, J.T., Unsworth, B.A. and Preece, T. (1982). A note on ginger blotch, a new disease of the cultivated mushroom *Agaricus bisporus*, *J. Appl. Bacteriol.*, 52: 43–48.

Wood, F.C. (1950). Bacterial rot and weeping disease of the cultivated mushroom, *Mushroom News*, 3: 23–27.

Wuest, P.J. and Zarkower, P.A. (1991). Mummy disease of button mushrooms. Causation, crop loss, mycosphere implications, In: *Science and Cultivation of Edible Fungi,* M.J. Mather (Ed.). A.A. Balkema Publishers, Brookfield, USA.

Yamanaka, K. (2006). In Chang, S.T. (2007). Development of world mushroom industry and its roles in human health. In: *Mushroom Biology and Biotechnology*, R.D. Rai, S.K. Singh, M.C. Yadav and R.P. Tewari (Eds.), Mushroom Society of India, Chambaghat, Solan, India.

Yang, X.M. (1986). *Cultivation of Edible Mushroom in China*, Agriculture Printing House, Beijing, Peoples Republic of China, pp. 489–510.

Yau, C.K. and Chang, S.T. (1972). *Cotton Waste for Indoor Cultivation of Straw Mushroom*, World Crops, 24: 302–303.

Ying et al., 1987 (reference cited in Sharma, V.P. Kumar, S. and Tewari, R.P. (2009). *Flammulina velutipes , The Culinary Medicinal Winter Mushroom, Tech. Bull.*, Directorate of Mushroom Research, Chambaghat, Solan. p. 53.

Yoshioka, Y., Sano, T. and Ikekawa, T. (1973). Studies on antitumor polysaccharides of *Flammulina velutipes* (Curt. Ex Fr.) Sing., *I. Chem. Pharm. Bull. 21*: 1772–1778.

Young, J.M. (1970). Drippy gill: A bacterial disease of cultivated mushrooms caused by *Pseudomonas agarici* n.sp., *J. Agric. Res.*, New Zealand, 13: 977–990.

Zadrazil, F. (1975). Influence of CO_2 concentration on the mycelium growth of three *Pleurotus* species, *Eur. J. Appl. Microbiol.*, 1: 327–335.

____ (1978). Cultivation *Pleurotus,* pp. 521–557, *The Biology and Cultivation of Edible Mushrooms*. In: S.T. Chang and Hayes, W.A. (Eds.), Academic Press, New York, p. 819.

Zadrazil, F. and Schliemann, J. (1975). Ein Beitrag zur Okologie und Anbautechnik von *Stropharia rugoso-annulata*/Farlow ex Murr. Champignon, 163: 7–22.

Zeng, Y.H. and Xi, Y.F. (1994). Preliminary study on colour fixation and controlled atmosphere storage of fresh mushrooms, *Journal of Zhejiang Agricultural University*, 20(2): 165–168.

Index